SEISMIC DESIGN
CODES AND PROCEDURES

Glen V. Berg

The University of Michigan
Ann Arbor

EARTHQUAKE ENGINEERING RESEARCH INSTITUTE

Published by

The Earthquake Engineering Research Institute, whose objectives are the advancement of the science and practice of earthquake engineering and the solution of national earthquake engineering problems.

This is volume six of a series titled: Engineering Monographs on Earthquake Criteria, Structural Design, and Strong Motion Records.

The publication of this monograph was supported by a grant from the *National Science Foundation*.

Library of Congress Catalog Card Number 83-80263
ISBN 0-943198-25-9

This monograph may be obtained from:
Earthquake Engineering Research Institute
2620 Telegraph Avenue
Berkeley, California 94704

FOREWORD

The occurrence of earthquakes poses a hazard to urban and rural areas that can lead to disaster unless appropriate engineering countermeasures are employed. To provide an adequate degree of safety at an affordable cost requires a high level of expertise in earthquake engineering and this in turn requires an extensive knowledge of the properties of strong earthquakes, the influences of soil conditions on ground motions, and the dynamics of structures that are moved by ground shaking. To achieve this it is necessary for relevant information to be published in an appropriate form.

This monograph by G.V. Berg on the development of seismic design codes and procedures is the sixth in a series of monographs on different aspects of earthquake engineering. The monographs are by experts especially qualified to prepare expositions of the subjects. Each monograph covers a single topic, with more thorough treatment than would be given to it in a textbook on earthquake engineering. The monograph series grew out of the seminars on earthquake engineering that were organized by the Earthquake Engineering Research Institute and presented to some 2,000 engineers. The seminars were given in 8 localities which had requested them: Los Angeles, San Francisco, Chicago, Washington D.C., Seattle, St. Louis, Mayaguez P.R., and Houston. The seminars were aimed at acquainting engineers, building officials and members of government agencies with the basics of earthquake engineering. In the course of these seminars it became apparent that a more detailed written presentation would be of value to those wishing to study earthquake engineering, and this led to the monograph project. The present monograph discusses the reactions of structures to earthquakes, the development of seismic building codes, and the seismic provisions of the major building codes now in use and the tentative provisions that will effect changes in the coming years.

The EERI monograph project, and also the seminar series, was supported by the National Science Foundation. EERI member M. S. Agbabian served as Coordinator of the seminar

5

series and is Coordinating Editor of the monograph project. Technical editor for the series is J. W. Athey. Each monograph is reviewed by the members of the Monograph Committee: M.S. Agbabian, G.V. Berg, R.W. Clough, H.J. Degenkolb, G.W. Housner, and C.W. Pinkham, with the objective of maintaining a high standard of presentation.

Monograph Committee
November, 1982

PREFACE

Building codes go far back in civilization. The code of laws of Hammurabi, king of Babylonia about 2100 B.C., included the earliest known legal requirements for buildings. The code gave no guidance about building a house, but it imposed penalties on the builder if things went wrong, including the death of the builder if the house fell and killed the owner. Modern building codes in the United States set out rules governing the design and construction of buildings, intended to reduce to an acceptably low level the likelihood of building failure. The goal is the same as that of Hammurabi; the means are different.

This monograph is intended to introduce the reader to the design of earthquake-resistant structures. It considers first what causes earthquakes and how structures react to them, next the development of seismic building codes, then the seismic provisions of the major building codes now in use in the United States, followed by new tentative provisions for the development of seismic building codes, recently published but not yet enacted, which may guide the changes in building codes in the United States over the next several years. Finally, it takes a look at some structures that have gone through destructive earthquakes, to see how they have responded and what characteristics of their design and construction have led to good or bad performance.

Seismic building codes are relatively recent, largely because we have only recently gained an understanding of the nature of earthquake forces and their effects on structures. Seismic building codes are transient. Some of their provisions last for many decades; others are revised or replaced after only a year or two. We give here a snapshot of current codes and procedures. The principles are durable; the details may be ephemeral.

The comments of Henry J. Degenkolb and Clarkson W. Pinkham on the technical content of the manuscript and the style editing of Margaret E. Berg are sincerely appreciated.

Glen V. Berg
The University of Michigan

7

TABLE OF CONTENTS

PAGE

Introduction 11
    Theory of Vibrations 15
    Behavior of Structures............................. 16

Development of Seismic Building Codes 19
    American Code Development 24
    Building Code Objectives 28

Current American Building Codes..................... 33
    BOCA, National, and Standard
        Codes and ANSI-72............................. 35
    Uniform Building Code............................. 46
    The Design Process and Some Notes
        of Caution..................................... 57

Applied Technology Council Provisions................ 61
    Seismic Performance Categories 62
    Equivalent Lateral Force Procedure 69

Code Comparison 82
    Comparative Requirements in
        Maximum Seismic Zone........................ 82
    Comparative Requirements in Buffalo 86

Experience in Past Earthquakes...................... 88

References 118

RUAUMOKO

Maori god of volcanoes and earthquakes, New Zealand

Seismic Design Codes and Procedures

by
Glen V. Berg

INTRODUCTION

Among the forces of disaster in nature the earthquake is unique in occurring without any consistently recognizable warning. Premonitory signals may occur, and research toward learning what they are is proceeding. However, to date there have been very few successful predictions of destructive earthquakes. One outstanding success was achieved in 1975 at Hai Cheng, China, where a timely prediction made possible the evacuation of the city several hours ahead of the event, saving perhaps thousands of lives. Remarkable as it was, that success did not herald a solution to the problem, for not only have a few subsequent destructive earthquakes in China gone unpredicted, other recent predictions have proved false. Lacking a suitable warning system, we must continue to rely heavily upon protective construction of buildings to reduce the adverse effects of earthquakes. Where we are unsuccessful, we can only try, after the disaster, to alleviate the human suffering and economic loss through humanitarian response.

Surely no natural catastrophe can be more horrifying than an intense earthquake. Consider the beautiful seaport city of Agadir, Morocco. On the evening of February 29, 1960, a festive crowd was attending a Leap Year party in the Saada Hotel. The earthquake came, and in less than a minute the hotel shown in Fig. 1 was reduced to the pile of rubble shown in Fig. 2. In that one building 400 persons died, and there were 12,000 deaths in the city of Agadir (Ref. 1). This was not a great earthquake in seismological terms, for earthquakes as large on the Richter magnitude scale occur somewhere on earth more often than once a month on the average. However, most of them occur in sparsely populated regions or under the oceans, and few besides seismologists take note. Agadir was unlucky.

Courtesy of American Iron and Steel Institute

Figure 1. Saada Hotel, Agadir, Morocco, before the earthquake.

Courtesy of American Iron and Steel Institute

Figure 2. Collapsed Saada Hotel, Agadir.

What causes earthquakes? All through history, man has sought an answer to that question. The Algonquin Indians of North America blamed the movements of a giant tortoise that they thought carried their land on its back. The Maoris of New Zealand attributed them to their earthquake god, Ruaumoko. The ancient Japanese believed the stirrings of a giant catfish deep in the ocean made their island tremble. In Western culture many believed the earthquake was the punishment of a wrathful God for man's sinful ways.

Aristotle made an early attempt at a scientific explanation of earthquakes. He theorized that winds became entrapped in underground caverns and caused the earth to tremble when they tried to escape. That explanation was consistent with the known locations of earthquakes, which were for the most part adjacent to the oceans, where winds might enter underground passages at low tide and then get trapped when the tide rose. Aristotle's theory persisted for nearly two millennia. Shakespeare called upon it in *Henry IV*, in the passage where Hotspur rebukes Glendower's boast with the words:

> O, then the earth shook to see the heavens on fire,
> And not in fear of your nativity.
> Diseased nature oftentimes breaks forth
> In strange eruptions; oft the teeming earth
> Is with a kind of colic pinch'd and vex'd
> By the imprisoning of unruly wind
> Within her womb; which, for enlargement striving,
> Shakes the old beldam earth, and topples down
> Steeples and moss-grown towers. At your birth
> Our grandam earth, having this distemperature,
> In passion shook.

The fracture of crustal rock was finally recognized in the 1890's to be the real cause of earthquakes. Bunjiro Kotō, Professor of Geology at Tokyo Imperial University, first made the crucial association. Ground breakage in earthquakes had of course been observed earlier, but it had always been seen as an effect rather than the cause. Harry Fielding Reid extended Kotō's idea after the San Francisco earthquake of 1906. There, the surface effects of the earthquake included offsets as great as twenty feet along the San Andreas fault. On the basis of such observations, along with the records of earlier triangulation surveys in the vicinity of the fault, Reid theorized that the crustal rock in the vicinity of the fault had undergone a gradual strain over a period of years until finally the stress overcame the strength of the rock at some location, whereupon the rock fractured anew and slipped back toward a stress-free state. The tremendous amount of strain energy that had accumulated in the rock during the years of distortion was thus suddenly released, and propagated in all directions from the fault break as a series of shock waves felt throughout the region as the chaotic ground motion of the earthquake. That, in brief, is the Reid elastic rebound theory. Although it leaves a few observed phenomena unexplained, it is still today the most widely accepted theory of the earthquake mechanism.

Reid's theory does not say what distorts the crustal rock in the first place. Until a few decades ago, that remained a mystery. Now, with the benefit of hindsight, it is difficult to see how perception of the cause could have been so elusive. In 1915 Alfred Wegener, Professor of Meteorology and Geophysics at the University of Graz in Austria, first published his book on the origin of continents and oceans (Ref. 2), in which he asserted that the continents had once been united and had separated and moved to their present positions, driven by forces that he did not entirely explain, but that he thought were somehow associated with the earth's rotation. Wegener's theory of continental drift was largely rejected by the scientific community. It was not until the 1950's that Maurice Ewing and his colleagues made the sea floor studies that revived Wegener's theory and began to fill it out. Now, with the compelling evidence of more recent oceanographic and geologic studies, it seems abundantly clear that the continents were indeed once a single land mass, gradually drifting in huge crustal segments or plates to their present positions over a period of perhaps 150 million years.

While earthquakes occur all over the globe, most of them take place at the edges of the immense moving crustal plates. One major plate boundary follows along the west coast of North America, where the Pacific plate is being thrust underneath the American plate, thereby causing the earthquakes of California, and we have come to think of California and Alaska as being America's earthquake country. However, a look at the historical record reveals some surprises! Some of the strongest earthquakes ever to have occurred in the United States took place in 1811 and 1812 near New Madrid, a small town in the bootheel of Missouri. The New Madrid region was only sparsely populated, so the earthquakes were not catastrophic in human terms. Nonetheless, land movements were massive, including the formation of Reelfoot Lake, and the shaking was strong enough to topple chimneys in Cincinnati, 375 miles away. Strong earthquakes also struck Boston in 1755 and Charleston, South Carolina, in 1886. New Madrid, Boston, and Charleston are all far from the edge of any crustal plate.

Theory of Vibrations

The effect of an earthquake on a building is a shaking of the foundation beneath it. If the building has sufficient strength and resilience, it will move along with the ground and vibrate. If it is too weak or too brittle, damage and possibly collapse will ensue. The aim of earthquake engineering is to ensure to the extent possible that the necessary strength and resilience will be present in the completed building.

At first glance, earthquake-resistant design might seem to be a straightforward engineering application of the theory of structures and the theory of vibrations. But while the theories are well established, they are difficult to apply. Another monograph in this series considers the behavior of a structure in an earthquake from an analytical point of view (Ref. 3). It presents mathematical methods that would enable one to compute the response of a structure to an arbitrary input motion. However, while such analyses necessarily are idealizations, structures and earthquakes are real.

One problem is the earthquake itself. Earthquake motion is chaotic and sometimes violent, and it involves translation and rotation of the ground in all directions simultaneously. Strong motion seismographs record the translational components and we have many records of them. But up to now the rotational components have received little study. They are generally dismissed as being of minor consequence. Possibly the extreme difficulty of recording them provides additional reason for their neglect. Except for the rotational components, we believe we understand the motion of recorded strong earthquakes reasonably well.

Behavior of Structures

A second problem lies in the behavior of the structure. In dynamic analysis of structures, we model the structure mathematically in well-defined although often cumbersome formulations. Elastic behavior is assumed sometimes in research and nearly always in practice, which leads to equations of motion that can be solved readily either as a system of coupled linear differential equations or, with normal modes procedures, as uncoupled equations. Nonlinear behavior gives rise to difficulties, for it invalidates the normal modes procedure; however, if the equations are not ill conditioned, they are still solvable as coupled equations. Alternatively, if the nonlinearity is not severe, the behavior of the system can sometimes be approximated by linear behavior over a limited range.

While individual structural members are ordinarily elastic, or very nearly so, if deformations remain small, connections respond to the forces exerted on them in complex ways that are quite difficult to define mathematically. Complicated paths of force transfer lead to stress concentrations. Moreover, the cutting, forming, and welding processes leave high residual stresses. For both of these reasons, the yield stress ordinarily is reached in the joints or connections well before the structural members yield. While partial yielding does not reduce the strength of the connection, it does affect its stiffness. Hence the behavior of a structure cannot be inferred accurately from the behavior of the members alone. A greater degree of uncertainty is likely to be introduced by the more complex parts of the structure, such as the stairways, stairwell and elevator enclosures, exterior walls, and mechanical equipment. The in-plane stiffness of walls and partitions is particularly important. Partitions may be considered nonstructural components by the designer, but simply designating them as nonstructural on the plans does not make them behave that way. With sufficient care in detailing, they may be isolated so that they do not contribute significantly to the stiffness of the structure.

The mass of the nonstructural components can be taken into account quite readily. However, to determine the damping attributable to even bare structural members and components is difficult, and to appraise accurately the influence of the nonstructural components on damping is virtually impossible. The foundation, too, affects the dynamic behavior of the structure, and to represent faithfully the influence of soil-structure interaction is troublesome, to say the least.

If we could overcome all of the foregoing difficulties so that all six components of the input motion were known, and all of the inertia, damping, and stiffness properties were defined fully and exactly, it would be possible, at least in theory, to calculate the response to any desired degree of precision. In fact, however, such calculations would be enormously expensive and time consuming, even with the best of computing facilities. Many simplifications must be introduced, even for research purposes, and for design the costs would otherwise quickly get out of hand.

It is a sobering thought to recognize that not even the most sophisticated computer programs for structural analysis can calculate the stresses due to dead load, but upon reflection the reason becomes apparent. The analysis treats the structure as though it were built to its exact dimensions in a gravity-free environment and then, after completion of construction, gravity loads were applied to the stress-free structure. Effects thus calculated can be vastly different from those actually encountered by the structure. For example, if we had a very tall building, say forty or fifty stories, with very large interior columns and very small peripheral columns, the peripheral columns would be more heavily stressed by dead load than the interior columns would be, and they would therefore shorten more. In the mathematical model, this differential column shortening would accumulate over the height of the building, leading to large calculated dead-load bending moments in the peripheral columns of the top story. On the other hand, the dead load of the real structure would be applied one floor at a time, as construction progressed. There would be differential column shortening, to be sure, but it would not accumulate over the height of the building. Each floor would be constructed level, thus compensating for any differential column shortening in the stories below. Thus the dead-load bending moments in the columns of the top story would be nearly zero.

There is yet the ultimate difficulty: the structure must be able to resist not the earthquake that occurred at some instrumental location years ago, but earthquakes that may occur at the site of the building during the twenty or fifty or one hundred years of its intended useful life. We can reasonably presume that the characteristics of future earthquakes will be similar to those of the past, and surely our design procedures should be consistent with and firmly based on the principles of structural mechanics. However, they will involve many uncertainties and they must incorporate many factors of engineering judgment.

Civil engineering structures are basically different from other engineering structures. The automobile designer or the aircraft designer produces a design that will be used for hundreds or thousands of identical structures. For them it is cost effective to test models or components extensively and then to test full-scale structures, and to modify, re-test, and remodify until the design is finally honed down to the production model. On the other hand, civil engineering structures — bridges, dams, tall buildings — are ordinarily one of a kind, and there is no opportunity to improve the design or correct errors after that one is built. Unlike the airplane, the civil engineering structure has to fly the first time.

THE DEVELOPMENT OF SEISMIC BUILDING CODES

A few types of buildings of centuries past have proved remarkably resistant to earthquake forces. For example, wooden pagodas in Japan constructed before the fifteenth century, some of them among the tallest wooden structures in the world, have survived earthquake after earthquake with never a report of serious damage (Ref. 4). Pagodas are relatively flexible structures, having natural periods in the range of 1 to 1.5 seconds, considerably longer than most other structures in Japan and longer than the dominant period of ground motion in Japanese earthquakes. Wooden structures are relatively light in weight, and hence incur smaller inertia forces than some other types of structures. However, the remarkable ability of Japanese pagodas to withstand earthquakes must be attributed largely to their structural damping, for any deformation of a pagoda is accompanied by the friction of timber sliding on timber and wood on wood in the contact surfaces of timber joints.

The satisfactory behavior of these ancient Japanese structures in earthquakes was achieved more by coincidence than by design. The design of structures specifically to withstand earthquake damage is a recent development and, as one would expect, has come about largely in response to specific disasters. Three events of great influence were the San Francisco earthquake of 1906, the Messina-Reggio earthquake of 1908, and the Tokyo earthquake of 1923. John R. Freeman discusses them in his important 1932 book *Earthquake Damage and Earthquake Insurance* (Ref. 5).

The San Francisco earthquake was the greatest of the three by seismological measures, although in disastrous results it was far surpassed by the other two. It was caused by a break more than 200 miles long in the San Andreas fault, with relative displacement on opposite sides of the fault reaching 15 or 16 feet — even 21 feet at one location, where ground lurching may have contributed to the offset. In the city of San Francisco, located several miles from the fault, some 700 persons were killed and property loss reached $400 million. Fire, rather than the shaking itself, was the great destroyer of property. Ground displacements severed water mains, so that fires could not be fought effectively. They burned out of control for three days. Teams of engineers who investigated the relative proportions of earthquake and fire damage in the city reported that about 80 to 95% of the damage in the burned-out district was caused by fire, and only 5 to 20% by shaking. There are some indications that shaking may have been more severe at certain other locations, Stanford University and San Jose in particular.

The San Francisco earthquake demonstrated that some good buildings of the day were quite capable of withstanding earthquake shaking. Wood frame buildings responded exceptionally well. Indeed, Freeman reported that the majority of the buildings in the city up to five stories high that were well designed and well constructed performed satisfactorily except for those built on soft ground or fill. He concluded that for tall buildings, the safest structural system was a steel frame made as rigid as practicable by large gusset plates connecting steel columns to steel floor beams and spandrel girders, all embedded in monolithic walls of reinforced concrete and connected and braced horizontally by rigid reinforced concrete floors and roof.

Although the 1906 earthquake awakened American engineers to the need for earthquake-resistant design and inspired changes in construction requirements, the appearance of explicit earthquake lateral force provisions in American building codes would await the Santa Barbara disaster nearly twenty years lateer.

In Italy, a land of frequent earthquakes, development of earthquake-resistant construction began following the Calabrian earthquakes of 1783. On the basis of a comparative study of the buildings that had survived the earthquakes and those that had failed, the Italians developed a new type of structure consisting of a timber frame infilled with stone embedded in mortar. This advance was made unaided by engineering analysis, but the result was a structural system admirably suited to its purpose. Resistant to earthquake forces, it employed indigenous materials available to the people at low cost, and buildings could be built with simple tools by semi-skilled or unskilled labor. This type of construction is employed today in much of southern Europe, and variants appear all over the world, for example, the Taquezal construction of Nicaragua. Because the infilled timber frame is not suited to tall buildings, ordinances were enacted in 1784 setting a two-story height limit on all new buildings and requiring that any existing structures taller than two stories that had been damaged must be cut down to two stories. Alas, the ordinance was not enforced. The great Messina-Reggio earthquake of 1908 killed 160,000 persons in a relatively small area centered near the two cities on the Messina Straits. While the few infilled timber frame buildings withstood the shaking well, most of the buildings had been built of rubble masonry, and many were already weakened by previous earthquakes. Nearly all of these collapsed.

The scientific study of earthquake-resistant building practice in Italy was first undertaken following the Messina-Reggio catastrophe. At that time, a commission was appointed consisting of nine prominent practicing engineers and five distinguished university professors, and was charged with finding methods of designing buildings that would resist earthquakes, that could be erected easily, and that would be inexpensive enough to be within the reach of the devastated population. Two contradictory proposals emerged from the commission's deliberations. One favored isolating the building from the ground by a sand layer underneath the foundations or by supporting the building columns in the bottom story on spherical roller bearings that would permit horizontal movement. The other favored connecting the building firmly to a rigid foundation. The commission adopted the latter proposal. It proscribed unreinforced masonry houses taller than one story as well, and imposed other constraints on building systems. In 1909, basing its decision primarily on a study of three timber-framed buildings that had survived the Messina earthquake with little or no damage, the commission adopted a seismic ratio of $1/12$; that is, buildings had to be designed to withstand a lateral force of $1/12$ of their own weight. Three years later this was modified to provide that the ground story must resist a lateral force of $1/12$ of the weight above, and the second and third stories must resist $1/8$ of the weight above.

The remarkable accomplishments of the 1908-09 Italian commission set a new direction for earthquake-resistant design in Italy and opened a new era of earthquake engineering research there. For the class of structures it considered, the commission had arrived at recommendations that remain as valid today as when they were formulated more than seventy years ago.

Japan has endured as much earthquake destruction as any nation in the world. Among its most severe earthquakes was the one that devastated Tokyo and Yokohama on September 1, 1923. Earlier severe earthquakes had occurred in Tokyo, notably those of 1649, 1703, and 1855; but the 1923 event was far more destructive to that city than any of its predecessors. Altogether, it destroyed more than 140,000 lives and damaged property in excess of two billion dollars. Fire was the major cause of the property damage, accounting for about 90% of the damage in Tokyo and Yokohama. Indeed, a striking aspect of the 1923 event was the low degree of shaking damage suffered by well-designed structures.

Because of earlier strong earthquakes, Japanese architects and engineers had necessarily been alert to the need for earthquake-resistant structural design long before 1923. The disastrous Mino-Owari earthquake of 1891 had provided them with strong impetus to design structures capable of withstanding heavy shaking. They had especially studied strong ground motion, and had established an active seismological program about 1880. By 1900 Professor Fusakichi Omori had developed the Omori scale of earthquake intensity relating ground acceleration to the degree of damage done to both natural and man-made structures. Lacking strong motion accelerograms, which would remain unavailable for many more years, Professor Omori obtained the accelerations in his scale from shaking-table experiments and from field observations of fallen columns and fallen stone lanterns.

Well before the 1923 earthquake, Japanese architects and builders had begun to use reinforced concrete and structural steel, and structures built of those materials withstood the shaking well. Indeed, an engineering survey disclosed that only about 10% of the reinforced concrete buildings in the city of Tokyo had been severely damaged and nearly 80% were undamaged.

The two conflicting ideas about the merits of building flexibility noted earlier among the Italian engineers appeared in Japan as well. One engineering investigation, which included an extensive

22

study of the dynamic behavior of simple structures as well as an evaluation of the damage incurred in the 1923 earthquake, reported that masonry structures performed the worst and reinforced concrete next, and that steel and wood were the most reliable. That study reported that rigid structures were unreliable for resisting earthquakes, although other engineering investigations drew opposite conclusions from the same evidence. In Japan, as in Italy, the proponents of rigidity prevailed, and among them was Dr. Tachu Naito, then Professor of Architecture at Waseda University in Tokyo.

In 1923, Dr. Naito was already among the eminent engineers of Japan and his standing in the architectural and engineering communities was greatly enhanced by the performance of three of his buildings in the 1923 earthquake. He had designed his Japan Industrial Bank, Jitsugyo Building, and Kabuki Theater in Tokyo to resist a lateral force equal to 1/15 of their weight, and all three came through the earthquake virtually unscathed. The Japan Industrial Bank was a steel frame structure, the Jitsugyo Building a reinforced concrete frame, and the Kabuki Theater a combination of the two. All three had been generously endowed with braced bents and shear walls and were therefore quite rigid. By contrast, two other buildings, the Yusen and Marinochi buildings, were quite flexible. They, too, had been designed to resist lateral forces, but they suffered such extensive nonstructural damage that the cost of repair was about a third of the original cost of the building. The Japan Industrial Bank and the Yusen and Marinochi buildings were all 100 ft. high, at that time the legal limit for height.

Dr. Naito proposed four fundamental principles of earthquake-resistant design: First, a building should be designed to act as much like a rigid solid body as conditions would permit. To this end, structural members should be rigidly connected and generously braced. Dr. Naito saw this as a way to keep building periods short and thereby prevent resonance with ground motion. Second, a closed plan layout should be used; that is, the plan shape of the building should be a complete closed rectangle rather than a U, L, T, or H shape. Third, rigid walls should be used abundantly and disposed symmetrically in plan, and they should be continuous over the height of the building. Fourth, lateral forces should be allocated to the bents of the building in accordance with their rigidities. He developed a modified portal method of frame analysis that would permit rapid and reasonably accurate analysis of a bent to determine its rigidity.

The 1923 earthquake led the Home Office of Japan to adopt a number of changes in building regulations. A seismic coefficient of 1/10 was prescribed for all important new structures; that is, such structures had to be designed to have adequate strength at any level to withstand a horizontal force of 1/10 of the weight above. In practice, many of the more conservative designers or building owners used even larger seismic coefficients. In addition, size limits were made more restrictive for the rebuilding of Tokyo and Yokohama. The earlier height limit of 100 ft. above street level was retained.

Any discussion of the 1923 Tokyo earthquake for American readers would be incomplete if it failed to mention Frank Lloyd Wright's famous Imperial Hotel, which survived with considerably less damage than other buildings in the area. It was a most unusual building, with an area of about 300 by 500 ft. in plan, and built in a number of similar sections. Throughout most of the building there were two stories and a basement. The flat slab floors were supported on columns and exterior walls. Because the soil was very soft and ground water was only a few feet below the surface, Wright used short, closely spaced piles a mere 8 ft. long on a 2 by 2 ft. grid for his foundations, a feature to which he ascribed much of the credit for the favorable response of the building. A construction feature that received less comment from the architect, but which may have been a greater contributor to the successful performance of the building, was the exterior wall construction — two separated wythes of brick with the void space between them carefully filled with poured concrete. In principle, this is not greatly different from the filled-cavity masonry walls required for California schools today.

American Code Development

Although one might have expected intense American activity toward the enactment of seismic building regulations following the San Francisco earthquake of 1906, it did not occur. The code governing reconstruction of the city of San Francisco stipulated a 30 psf wind force, which was intended to protect against both wind and earthquake damage. Not until the 1925 Santa Barbara earthquake was serious and continuing legislative effort directed specifically toward the enactment of a seismic building code. In that year Congress assigned to the U.S. Coast and Geodetic Survey the responsibility for studying and reporting on strong motion seismology. The Pacific Coast Building Officials Conference (which later became the International Conference of Building Officials) published in 1927 the first seismic design provisions in any of the major regional American building codes, in its *Uniform Building Code*, first edition. The provisions were contained only in an appendix and were not mandatory. Several California municipalities, however, did adopt mandatory earthquake provisions in their building codes, and there was an effort among California civic bodies, particularly the California State Chamber of Commerce and the Commonwealth Club of California, to get earthquake provisions enacted more widely. The 1933 Long Beach earthquake provided an unwelcome nudge.

School buildings were hit hard in the Long Beach earthquake, and Americans, then as now, got highly aroused about school safety. Such tragedies as the collapse of hotel walkways or the crash of a passenger airplane costing scores of lives fail to provoke the vigorous and lasting outcry that a tragedy causing death or injury to school children will bring. Perhaps that augurs well for the future of America. At any rate, spurred by the school building failures in Long Beach, the California State Legislature quickly enacted the Field Act, which makes the State of California Division of Architecture responsible for approving public school plans and supervising public school construction. The rules and regulations adopted by the Division of Architecture required that masonry buildings without frames be designed to resist a lateral force equal to 10% of the sum of dead load and a fraction of the design live load. For other buildings the lateral force was set at 2% to 5%, depending upon the allowable foundation loads. Shortly thereafter the Riley Act was adopted by the State of California, establishing a mandatory seismic design coefficient of 2% of the sum of dead load and live load for most buildings in the state, making exceptions only for rural dwellings and buildings not intended for human occupancy. This measure stands as the first statewide requirement for earthquake-resistant design enacted within the United States.

The City of Los Angeles next imposed a lateral force requirement of 8% of the sum of dead load and half of the design live load. This was also adopted in 1935 by the Uniform Building Code for Zone 3, its zone of highest seismicity.

In 1943, in a new departure, the City of Los Angeles enacted a code that related the seismic coefficient to the flexibility of the building — the first such code in the United States and among the first anywhere. The code established the relationship indirectly, by stipulating a seismic coefficient that varied among the stories of a building. The design lateral shear within each story was a seismic coefficient C times the dead load above. The coefficient C was stipulated to be $C = .60/ (N + 4.5)$, N being the number of stories above. Thus the base shear, that is, the total lateral force, ranged from .133 times the dead load for a one-story building ($N = 0$) to .0364 times the dead load for a 13-story building ($N = 12$), the maximum allowable building height. A later modification of the code changed the formula for C and removed the 13-story height limit.

The City of San Francisco had only the Riley Act provisions in its code until 1947, at which time it adopted lateral force requirements ranging from 3.7 to 8% of the design vertical load, the percentage depending on the number of stories, with variations for soil conditions. Then, in 1948, a Joint Committee of the Structural Engineers Association of Northern California and the San Francisco Section of the American Society of Civil Engineers was formed to draft model lateral force provisions for California building codes. By that time the strong motion accelerograph had been developed and a few strong motion records had been obtained at moderately close range. In addition, M. A. Biot had invented his earthquake response spectrum and his mechanical analyzer, and spectra had been evaluated for several strong motion records. Although the available information about strong motion was meager, it far surpassed anything that had been available for earlier code development.

The Joint Committee recommended that buildings be designed

for a base shear $V = C\,W$, in which W was the dead load plus one-quarter of the design live load, and the seismic coefficient was $C = .015/T$, T being the fundamental period of the building. This base shear was to be distributed along the height of the building as lateral forces at the floor levels, much as it is in current American codes. An empirical formula for the period T was given as $T = .05h/\sqrt{D}$, where h and D are the height of the building and its plan length in the direction being considered, both in feet. The City of San Francisco enacted a building code based on the Joint Committee recommendations in 1956, but with the seismic coefficient increased one-third to $C = .02/T$, with limits $.035 \leq C \leq .075$. Building period had now become an explicit factor in the determination of seismic design forces for the first time.

In 1960 the State of California requirements became more or less uniform. The Structural Engineers Association of California incorporated the lateral force recommendations of its Seismology Committee into the provisions of its building code, the SEAOC code. The provisions were adopted by many local building authorities in California and were also incorporated into the 1961 edition of the Uniform Building Code. They were nearly the same as the lateral force provisions in four of the current American codes, which will be discussed below.

Over the past few decades a pattern of American seismic building code development has emerged. Provisions have been adopted first by the Structural Engineers Association of California in its SEAOC code, then by the International Conference of Building Officials in the Uniform Building Code, and finally by the other major regional codes. Modifications have been made by SEAOC from time to time as new research findings have become available and as new experience has been gained from real earthquakes, fostering or inhibiting design or construction practices according to observed performance, good or bad. Code writing remains an art rather than a science, but it has become an art with a firm scientific base.

In a departure from the historical pattern of gradual and evolutionary code development, the Applied Technology Council, a research and development subsidiary of the Structural Engineers Association of California, embarked upon a code development project in 1974 under the sponsorhip of the National Science

Foundation and monitored by the National Bureau of Standards. Their intention was to take a fresh look at earthquake-resistant design in all of its aspects and to develop a new code suitable for adoption nationwide. Project participants came from all of the diverse specialties related to seismology and earthquake engineering — from engineering design offices, universities, industries, and government agencies all across North America. Nearly a hundred scientists and engineers contributed to the project report, *Tentative Provisions for the Development of Seismic Regulations for Buildings*, published jointly in 1978 by the Applied Technology Council, the National Science Foundation, and the National Bureau of Standards (Ref. 6). The report, not a code in itself, is intended as a guide for regulatory bodies in their development of building regulations. It is currently (1982) being reviewed by engineers and building officials to evaluate its utility, cost, and effectiveness in leading to earthquake-resistant structures. Some of the provisions of that report are presented in the sections that follow.

Building Code Objectives

The fundamental aim of earthquake engineering is to provide structures and facilities that are safe for public use even if subjected to that worst possible trial, the shaking of the earth beneath them. Economical structures are desirable, but the achievement of economy ranks far below public safety among code considerations. Somehow the knowledge gained from theory, analysis, research, and field observation, all needs to be translated into building regulations that are practical for use in the design office, that are enforceable by building officials, and that will lead to safe structures at a cost that society can afford. At the same time, the regulations should not inhibit the designer's creativity or hinder the development of new materials, new concepts, or new construction methods. Moreover, the regulations must be based on what we know now. Design decisions cannot await the arrival of new knowledge, however urgently needed.

The objective of a building code is well stated in Section 102 of the Uniform Building Code (Ref. 7):

The purpose of this Code is to provide minimum standards to safeguard life or limb, health, property, and public welfare by regulating and controlling the design, construction, quality of materials, use and occupancy, location and maintenance of all buildings and structures within this jurisdiction and certain equipment specifically regulated herein.

The Structural Engineers Association of California has a code philosophy that would probably be accepted by most engineers. The primary function of a building code, according to SEAOC, is to provide minimum standards to assure public safety. It should provide safeguards against major failure and loss of life. Damage limitation is to be sought but is not in itself a code objective. With regard to earthquake response, the aim of the code is to lead to structures that can resist minor earthquakes undamaged, resist moderate earthquakes without significant structural damage even though incurring nonstructural damage, and resist severe earthquakes without collapse.

Throughout the world almost without exception, seismic building codes provide for potential earthquake hazard by requiring that structures be designed to resist specified static lateral forces. In most codes, but not all, the vertical forces due to the earthquake are not considered. The rationale for this omission is the belief that normal gravity load provisions, coupled with normal factors of safety, ought to provide a margin of safety sufficient to accommodate the dynamic effects of the vertical acceleration of the earthquake. Lateral forces are quite a different matter, however, for the normal state of lateral acceleration is zero, in contrast with the normal state of vertical acceleration of one g, the acceleration of gravity.

The concept of an equivalent static lateral force is somewhat misleading and often misunderstood. Most engineers are accustomed to designing for static forces. We think in static terms, and we accept force equilibrium without question. The design loads we usually consider, such as gravity, wind, and snow loads, represent a mean of peak loads, and the likelihood of their ever being exceeded is not great. It is not surprising that we extend this line of thinking to earthquake-resistant design.

However, the static lateral design forces stipulated in most seismic building codes do not represent the peak dynamic forces likely to be exerted on the structure by an earthquake. The code forces are "equivalent" to the earthquake forces only in the sense that a structure designed to resist the code forces without overstress should be able — if the design is carefully executed to account for stress reversals, provide adequate member ductility, and provide connections of sufficient strength and resilience — to resist minor earthquakes without damage, resist moderate earthquakes without significant structural damage, and resist major earthquakes without collapse.

A good engineer will strive to limit structural damage to repairable damage, even in a major earthquake. Still, it should be recognized that a structure competently designed by a well-qualified engineer to comply with all provisions of the code and constructed with the best of materials and workmanship under close engineering supervision, will nonetheless be far overstressed in a major earthquake and can be expected to suffer structural as well as nonstructural damage. Limiting structural damage to what can be repaired requires skill and sound engineering judgment extending well beyond merely satisfying the stress limitations stipulated in the code.

Although seismic building codes are not all alike, most of them share features drawn from the theory of structural dynamics. If it employs an equivalent static lateral force, the code must stipulate the magnitude of the force and its distribution along the height of the structure. The total lateral force, the base shear, should logically take into account the seismicity of the region, the geology and soil conditions at the site, the dynamic properties of the structure, and the anticipated importance of the building and its occupancy, or the possible social consequences of failure.

The seismicity of a region usually is shown in the code by means of a zone map or seismic risk map. Often the zones have been determined by seismological or historical records of the greatest earthquakes of the past without regard to frequency of occurrence. In recent years some effort has been directed toward microzoning; that is, delineating in a small region, ordinarily an urban region, zones of different degrees of seismic risk. Such microzones might indicate the areas of least risk suitable for any occupancy and, at the other extreme, show high risk areas suitable only for parks or buildings not intended for human occupancy. Microzones are ordinarily determined on the basis of soil conditions, height of the water table, surface geology, and the locations of known faults. A few current European building regulations include microzone provisions. One of the few trials of such provisions in an actual strong earthquake occurred in Bucharest in 1977. The results did not show great promise, for the greatest damage occurred in the microzone of least risk (Ref. 8).

Site conditions are often accounted for by a coefficient assigned according to the firmness of the soil at the building site. Experience in destructive earthquakes has demonstrated convincingly that most buildings on soft ground incur a greater degree of damage than comparable buildings on firm ground, and that the latter in turn generally suffer greater damage than comparable buildings on rock, although the reverse may be true for very stiff buildings. Many codes specify a reduced base shear for rock sites and an increased base shear for soft soil sites. Soft, medium, and hard soils are often defined in terms of bearing capacity or shear wave velocity, but in some codes they are undefined and this consideration is left to the judgment of the designer or the building official. Some current American codes acknowledge the influence of site conditions, and others do not.

The properties of the structure that affect the base shear are its mass or weight, its natural period, and its degree of damping. Should the building be distorted beyond the elastic range, as it would almost certainly be in a destructive earthquake, the yield level of the members and the nature of their inelastic behavior would also have an influence. Some codes take such behavior beyond the elastic range into account indirectly in the lateral force provisions by means of a factor related to the type of structure. Other codes take it into account directly in the member design provisions.

Finally, some codes insert into the design requirements a consideration of the social importance of the building by means of a factor that increases the design base shear or imposes more stringent drift limits for buildings housing essential urban facilities — such as hospitals and fire stations — or those whose failure would endanger large numbers of occupants — such as public schools or large assembly halls. This doubtless should be an important consideration. It would also seem that intended useful life should merit attention in determining seismic design requirements, with short-lived or temporary buildings made subject to less stringent design requirements than long-lived or monumental buildings. However, current American codes do not take this factor into account. They treat all buildings as though they were to be retained indefinitely.

CURRENT AMERICAN BUILDING CODES

There are four major model building codes in use in the United States today:

> *The BOCA/Basic Building Code*, issued by Building Officials and Code Administrators International, Homewood, Illinois (Ref. 9)

> *The National Building Code*, of the American Insurance Association, New York (Ref. 10)

> *The Standard Building Code*, of the Southern Building Code Congress, Birmingham, Alabama (Ref. 11)

> *Uniform Building Code*, of the International Conference of Building Officials, Whittier, California (Ref. 7)

Their use is somewhat regional: The Uniform Building Code is used most extensively in the West, BOCA in the Midwest, Standard in the South, and National in the Northeast. The codes do not have jurisdictional boundaries, however. All of these documents stipulate design requirements based on a blend of theory, experiment, and empirical observation.

In addition to these codes there is Standard A58.1, *American National Standard Building Code Requirements for Minimum Design Loads in Buildings and Other Structures*, of the American National Standards Institute, Inc., New York (Ref. 12). For brevity we will refer to this as ANSI-72. It considers only loads, not materials or structural design. As this monograph goes to press, a 1982 edition of American National Standard A58.1 has just been published. Its seismic provisions differ significantly from those of the 1972 edition.

Also in the offing is a set of proposed seismic regulations developed by the Applied Technology Council, a research and development subsidiary of the Structural Engineers Association of California. They are discussed in the next chapter.

Historically, the seismic provisions of the four major current codes were first developed by the Structural Engineers Association of California, then incorporated into the Uniform Building Code in its next edition, followed by inclusion in the other regional codes in their subsequent editions. All four of these codes appear in new editions every three to six years, with interim supplements, although the seismic provisions are not necessarily revised for every new edition. Thus a year or two or three may pass from the time a new seismic provision is developed until it first appears in the Uniform Building Code, and a few more years before it finds its way into the other codes.

Just what is a building code and what is it intended to accomplish? A building code is a set of minimum legal requirements governing the design and construction of buildings within a particular geographical area or political jurisdiction. The requirements are expressed as a set of rules that are intended to achieve satisfactory performance of any completed building under the loadings inherent in its use or imposed by the forces of nature. The rules do not purport to be optimal; that is to say, the codes do not suggest the most economical way of accomplishing the desired performance. Indeed, they cannot even assure that it will be achieved at all. In particular, strict adherence to the code will not preclude earthquake damage, nor is that viewed as an economically obtainable objective. Survival from minor earthquakes undamaged, from moderate earthquakes without significant damage, and from major earthquakes without collapse would be considered satisfactory performance. It is, alas, entirely possible that a building could be designed and constructed in full compliance with the applicable building code, right down to the last turn of the nut, with the best of materials and superb workmanship, and still be damaged severely in an earthquake.

A building code provides formulas and guidelines, as a cookbook provides recipes and instructions. But just as the recipes and instructions of the cookbook are not enough to ensure tasty and nutritious meals, so the formulas and guidelines of the building code are not enough to ensure satisfactory structures. Successful results demand more — quality ingredients, a knowledge of their function and purpose, and skill in combining them.

BOCA, National, and Standard Codes and ANSI-72

The BOCA, National, and Standard codes and ANSI-72 are nearly identical in their earthquake provisions and can be considered together, with their significant differences pointed out along the way. All of them are based on earlier developments of the Structural Engineers Association of California. All stipulate the total lateral force, the base shear V, to be the product of four factors,

$$V = ZKCW$$

in which Z is a zone coefficient, K a structure coefficient, C a seismic or dynamics coefficient, and W the weight.

The weight W is normally the total dead load, but an exception is made for storage or warehouse occupancy, for which one-quarter of the design storage load must be included. Anchored loads such as partitions are to be included in the dead load whether they are permanently fixed or movable. Live load and snow load are not included.

The zone coefficient Z, related to the seismicity of the region, is specified by means of a zone map. Figure 3 shows the National Building Code zone map. The BOCA and ANSI-72 maps are the same as National, and so is the Standard map except that it lacks the tongue of Zone 2 extending down into California — a meaningless difference, since few if any California jurisdictions use these particular codes and standards. Three zones are indicated, and the coefficient Z has a value of 1 in Zone 3, 1/2 in Zone 2, and 1/4 in Zone 1.

The zones on these maps correspond roughly to estimates of the extreme ground shaking experienced in the region in the historical past, but without regard to the frequency of occurrence of that shaking. Thus coastal California, extreme southeastern Missouri, and the St. Lawrence region of New York state are all placed in Zone 3. They have all experienced roughly the same extreme degree of shaking in the past, and there is no basis for expecting that any extreme shaking ever to occur in the future

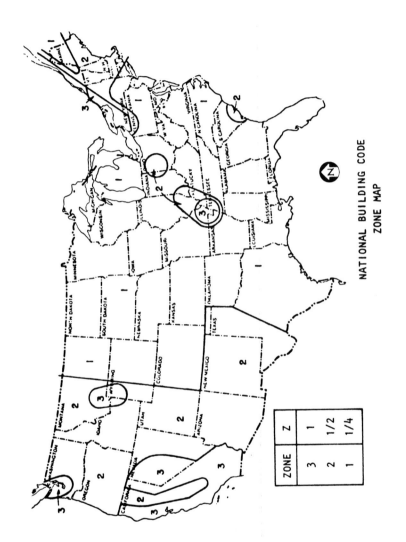

NATIONAL BUILDING CODE
ZONE MAP

ZONE	Z
3	1
2	1/2
1	1/4

Figure 3. National Building Code zone map.

would be greater in one of those locations than in another. However, frequency of occurrence of strong motion is another matter, for strong earthquakes have occurred far more often in coastal California than in the other locations. Thus, the probability of occurrence of motion of a given severity in any given time span may be far greater in one Zone 3 region than in the others.

The building coefficient K depends on the type of structure, ranging in value from 2/3 to 4/3 for buildings and on up to 3 for elevated tanks. The K factor, in effect, lowers the margin of reserve strength required for structural systems that have performed well in earthquakes, and raises the margin for systems that have performed badly. The complete set of K factors is given in Table 1.

Table 1

Building Coefficient, K

Structural system	K
Building with box system: No complete vertical load-carrying space frame; lateral forces resisted by shear walls.	1.33
Building with dual bracing system consisting of ductile moment-resisting space frame and shear walls, designed so that:	0.80
(1) Frames and shear walls resist total lateral force in accordance with their relative rigidities, considering the interaction of shear walls and frames.	
(2) Shear walls acting independently of space frame resist total required lateral force.	
(3) Ductile moment-resisting space frame has capacity to resist at least 25% of required lateral force.	

Table 1 (concluded)

Structural system	K
Building with ductile moment-resisting space frame designed to resist total required lateral force.	0.67
Other building framing systems.	1.00
Elevated tanks, plus full contents, on four or more cross-braced legs and not supported by a building.	3.00
Structures other than buildings.	2.00

The seismic coefficient C is given by the formula

$$C = \frac{.05}{\sqrt[3]{T}}$$

in which T is the natural period of the building. Figure 4 shows the relation, plotted with logarithmic scales. Log-log plots are convenient because they portray functions of this type as straight lines. A maximum value of 0.10 is stipulated, and C need not be taken to exceed 0.10, regardless of the building period. This appears in Fig. 4 as the horizontal branch for T < 0.125 sec. It is further stipulated that C shall be taken as 0.10 for all one- and two-story buildings.

The seismic coefficient C can be related to the response spectrum discussed by Anil Chopra in *Dynamics of Structures, A Primer* (Ref. 3), another monograph in this series. Figure 5 shows the 2% damping response spectrum curve for the Imperial Valley earthquake of May 18, 1940 — the notorious El Centro earthquake — superimposed on the seismic coefficient plot. This is the same response spectrum curve shown in Fig. 20 of *Dynamics of Structures*, replotted so that acceleration appears on the vertical logarithmic scale instead of a 45° inclined scale.

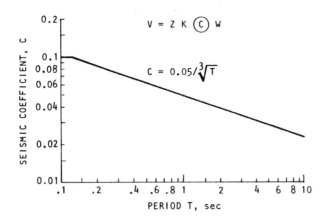

Figure 4. Seismic coefficient C, BOCA/National/Standard/ANSI-72.

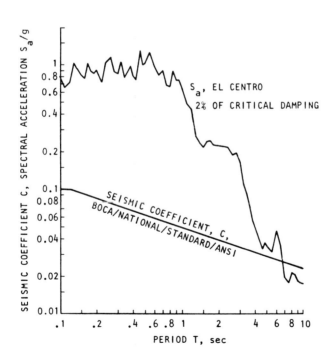

Figure 5. Code seismic coefficient and response spectrum compared.

The seismic coefficient C gives the design force as a fraction of the weight of the building, and the acceleration response spectrum S_a is the peak force on an oscillator as a fraction of the weight of the oscillator. Hence C and S_a have comparable but not identical meanings. A direct comparison is clouded by such issues as margins of safety on working stresses and the differences between static and dynamic structural behavior. Nevertheless, the spectral acceleration shown in Fig. 5 exceeds the seismic coefficient for nearly all periods, usually by a wide margin, by a factor of 20 at the extreme point. This is, of course, the spectrum for the El Centro earthquake, for many years the strongest earthquake motion ever recorded, and the damping ratio is low, only 2% of critical damping. Less intense ground motion or more damping would narrow the gap, but the spectral acceleration for a destructive earthquake could still exceed by a wide margin the seismic coefficient required by the code. This lends credence to earlier statements that damage may be expected even in well-designed structures in a major earthquake.

The seismic coefficient C is a function of building period, but how is the period to be determined? The Rayleigh or Stodola processes found in textbooks on structural dynamics could be used if the building stiffness were known, but that is not determined until the structure has been designed. We need the period in order to determine the base shear for which the structure is to be designed and the design of the structure will establish its stiffness, which we need in order to determine its period. The code provides a means of breaking the circle. It allows the period to be calculated either from "properly substantiated technical data," which could be a Rayleigh or Stodola process, or alternatively, from the empirical formulas

$$T = \begin{cases} 0.10 \, N & \text{for moment-resisting frame structures} \\ 0.05 \, h/\sqrt{D} & \text{for other buildings} \end{cases}$$

where
    N = Number of stories
    h = Height of building, ft.
    D = Depth in the direction being considered, ft.

These formulas express a design period in terms of building properties that are known at preliminary design stages. With the design period thus established, one can determine the seismic coefficient C, and then having Z, K, and W, one can find the design base shear V.

The code stipulates a force distribution, allocating the total base shear V as a set of lateral forces acting at the floor levels. A force F_t, computed according to the building dimensions, acts at roof level. The formula for F_t is:

$$F_t = \begin{cases} 0 & \text{for } h/D \leq 3 \\ 0.004 \ V \ (h/D)^2 & \text{for all other cases} \end{cases}$$

but in any case, $\quad F_t \leq 0.15 \ V$

The rest of the base shear is allocated to the floors in a triangular distribution, the force at each floor (including the roof) being proportional to the floor weight times its height above the base, thus:

$$F_x = (V - F_t) \frac{w_x h_x}{\sum w h}$$

In *Dynamics of Structures*, Chopra points out that the lateral force at any floor in any one mode of vibration is related to the base shear in that mode by the relation

$$f_j = \frac{V \ w_j \phi_j}{\sum w \phi}$$

41

Thus, if the mode shape is linear (where the displacement of each floor is proportional to its height above the base), then the code distribution would be the correct distribution of base shear for that mode. The linear shape is not an unreasonable approximation for the first mode shape of a tall building. If, on one hand, the building has relatively flexible columns and stiff girders, the mode shape is concave, as in Fig. 6(a). On the other hand, if the columns are relatively stiff and the girders flexible, the mode shape is convex, as in Fig. 6(b). The typical tall building is somewhere in between, and the linear shape of Fig. 6(c) used in the code is a convenient and reasonable approximation. Not all of the base shear is represented by the first mode, and for tall buildings where the higher modes may be significant, the extra force F_t at the roof is provided to account for them.

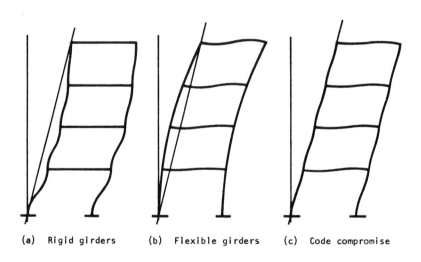

(a) Rigid girders   (b) Flexible girders   (c) Code compromise

Figure 6. First modes shapes.

The lateral forces on the building create not only shears in the stories, which induce shears and bending moments in the beams and columns, but also an overturning moment (OTM) which contributes to the axial forces in the columns, especially the exterior columns. These overturning forces augment the gravity load effect on the exterior columns on one side of the building and relieve it on the other side. If the overturning forces were great enough, they could cause distress in the exterior columns, either in compression or in tension. The code requires that the calculated overturning moment be considered, but it provides a reduction coefficient J for buildings of longer period.

$$\text{Base OTM} = J \sum F_i h_i$$

where
$$J = 0.6 \ / \ T^{2/3}$$

$$\text{but } 0.45 \leq J \leq 1$$

Figure 7 shows J as a function of period. The rationale for the reduction factor is that the base overturning moment is induced principally by the first mode of response, whereas the code forces are intended to represent all modes. In the first mode the lateral inertia forces at all floors act in the same direction, whereas in the higher modes some of the inertia forces act to the right while others act to the left. The higher modes therefore contribute less to the base overturning moment than does the first mode. If the higher modes are a significant part of the total response, which they are for tall buildings (i.e., long-period buildings), then the base overturning moment calculated from the code forces will be exaggerated. The reduction factor J is intended to offset the exaggeration.

Overturning moment in the stories above the base is similarly computed as the static moment due to the lateral forces above, but with a modified reduction factor:

$$J_x = J + (1 - J)(h_x \ / \ h_n)^3$$

where
$\quad J = $ Base OTM reduction factor
$\quad J_x = $ OTM reduction factor at level x
$\quad h_x, h_n = $ Height above base to level x, to roof

43

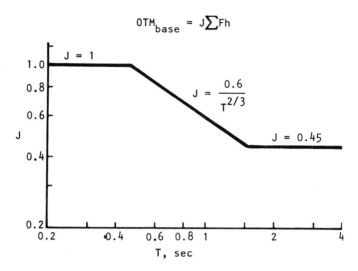

Figure 7. Base overturning moment factor J, BOCA/National/Standard/ANSI-72.

Thus J_x is equal to J at the base, nearly equal in the lower part of the structure, and increases to $J_x = 1$ at the top, as shown in Fig. 8.

Code stipulations also require that lateral displacement, or drift, be considered. The BOCA and Standard codes and ANSI-72 merely say that drift shall be considered in accordance with accepted engineering practice. The National Building Code quantifies its drift limitations, stipulating that if the height-to-depth ratio of the building exceeds 2.5, then the drift must not exceed one-half of one percent of the story height in any story.

The codes require that horizontal torsion as well must be taken into account. The inertia force induced by the earthquake acts through the center of mass, and if the structure is not dynamically symmetric, the center of rigidity and the center of mass may not coincide. The product of the story shear and the eccentricity of the center of mass above, measured from the center of rigidity of the story, would give a calculated torsion in the horizontal plane that must be considered in the design. Further, the code provides for an accidental torsion in each story, set at 5% of the product of

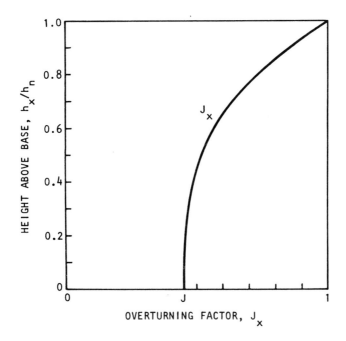

Figure 8. Variation of overturning moment factor J_x with height.

the story shear times the length of the building. This is to provide for any differences between the actual and calculated positions of the center of mass and center of rigidity. The 5% accidental torsion is a minimum total design torsion, not an addition to the calculated torsion. Note that the *length* of the building is used for accidental torsion, no matter whether earthquake forces are being considered for the long direction or the narrow direction.

Building setbacks may be troublesome because they produce abrupt changes in stiffness and mass. These four codes all stipulate that if the plan dimension of the tower in each direction is at least 75% of the corresponding plan dimension of the lower part, then the setback may be ignored in determining seismic forces. Otherwise, the tower is to be designed as a separate building, using the larger of two seismic coefficients at the base of the tower determined by considering the tower either (1) as a separate building for its own height, or (2) as a part of the overall structure. The resulting total shear at the base of the tower is then

45

to be applied as a lateral force at the top of the lower part of the building, which is to be designed separately for its own height.

It is required that all buildings more than 160 ft. high have ductile moment-resisting space frames capable of resisting at least 25% of the seismic forces for the structure as a whole, except that buildings more than 160 ft. high in Zone 1 may have shear walls or braced frames in lieu of a ductile moment-resisting space frame if they are designed with a horizontal force factor $K = 1.00$ or $K = 1.33$.

All of these codes use a working load basis, with steel structures designed according to the specifications of the American Institute of Steel Construction (Ref. 13), which use working stress procedures, and with reinforced concrete structures designed according to the specifications of the American Concrete Institute (Ref. 14), which use a load and resistance factor procedure. Overloads are permitted for load combinations that include seismic forces. The National code stipulates the following load combinations:

$$1.00 \times (\text{dead load} + \text{earthquake})$$
$$0.75 \times (\text{dead load} + \text{live load} + \text{earthquake})$$
$$0.75 \times (\text{dead load} + \text{earthquake} + \text{temperature})$$
$$0.66 \times (\text{dead} + \text{live} + \text{earthquake} + \text{temperature})$$

The design must be adequate for all of these combinations, with the earthquake force acting in any direction. These load combinations override other stress increase provisions, such as the AISC allowance of a 1/3 overstress in the presence of earthquake forces.

Uniform Building Code

The Uniform Building Code (UBC), which is used all along the Pacific coast and in many other locations as well, is based on the 1975 code of the Structural Engineers Association of California (SEAOC). It incorporates two additional factors, I and S, in the formula for base shear:

$$V = ZIKCSW$$

where, as in the other codes, W is the total dead load and, for storage or warehouse occupancy, 25% of the design storage load is added. Snow load is excluded if the design snow load is 30 psf or less. If the design snow load exceeds 30 psf, 25 to 100% of it is included, at the discretion of the building official.

The zone map, Fig. 9, has four zones instead of three — actually five, for there is a Zone 0 as well. The zone coefficient Z now has a value of 1 in Zone 4, $3/4$ in Zone 3, $3/8$ in Zone 2, and $3/16$ in Zone 1. No value is specified for Zone 0, but it seems unlikely that building officials in Zone 0 would require seismic design for their buildings except in unusual cases.

The importance factor I has to do with the degree of hazard to human life that failure of the building might cause, imposing greater design forces for some classes of occupancy than for others. For values of I, see Table 2.

Table 2

Importance Factor, I

Type of occupancy	I
Essential facilities, including hospitals and other medical facilities having surgery or emergency treatment areas, fire and police stations, and municipal government disaster operation and communication centers deemed to be vital in emergencies.	1.50
Any building where the primary occupancy is for assembly use for more than 300 persons in one room.	1.25
All others	1.00

The building coefficient K is the same as for the previous codes except that the factor for elevated tanks is reduced to 2.5 instead of 3.0. Values of K are presented in Table 3.

Table 3

Building Coefficient, K

Structural System	K
Building with box system: No complete vertical load-carrying space frame; lateral forces resisted by shear walls.	1.33
Building with dual bracing system consisting of ductile moment-resisting space frame and shear walls, designed so that: (1) Frames and shear walls resist total lateral force in accordance with their relative rigidities, considering the interaction of shear walls and frames. (2) Shear walls acting independently of space frame resist total required lateral force. (3) Ductile moment-resisting space frame has capacity to resist at least 25% of required lateral force.	0.80
Building with ductile moment-resisting space frame designed to resist total required lateral force.	0.67
Other building framing systems.	1.00
Elevated tanks, plus full contents, on four or more cross-braced legs and not supported by a building.	2.50
Structures other than buildings.	2.00

The UBC imposes a further requirement that is related to ductility but not reflected by the K-factor. For all buildings in Zones 3 and 4 and all those in Zone 2 having an importance factor

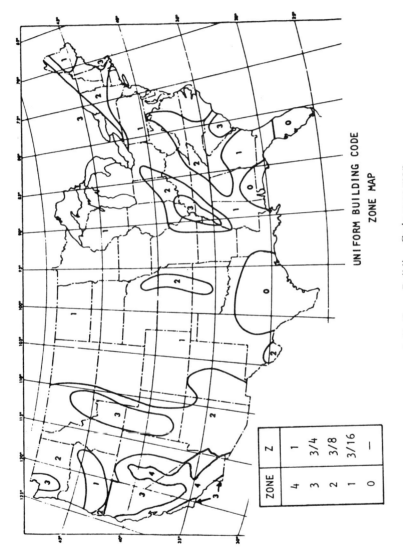

ZONE	Z
4	1
3	3/4
2	3/8
1	3/16
0	—

UNIFORM BUILDING CODE
ZONE MAP

Figure 9. Uniform Building Code zone map.

49

I greater than one, all members in braced frames must be designed for lateral forces 25% greater than those determined from the code provisions. Also, connections must either be designed to develop the full capacity of the members, or else be designed for the code forces without the one-third increase in stresses normally allowed when earthquake forces are included in the design.

The seismic coefficient C is specified as

$$C = \frac{1}{15\sqrt{T}}$$

but in any case C need not exceed 0.12. This is illustrated in Fig. 10. Unlike the codes discussed earlier, the UBC does not make an exception for one- and two-story buildings. The cutoff level for C at the short period end of the scale is 20% greater than for the codes discussed earlier.

Except for structures with periods longer than about 5.6 sec, the coefficient C is greater for the UBC than for the other codes. However, in Zones 3, 2, and 1, this is at least partially offset by the difference in the Z factor, which is 25% less for UBC. The Zone 4 region of UBC is contained in Zone 3 of the other codes, and for this region the Z factor is the same, Z = 1, for all of them.

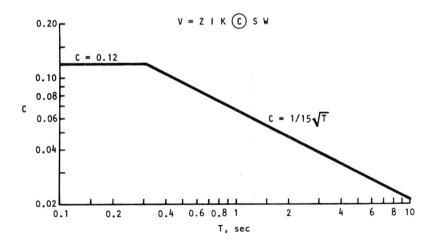

Figure 10. Seismic coefficient C, UBC.

The UBC provides a formula for period, based on the Rayleigh quotient, which can be derived by equating the peak kinetic energy to the peak strain energy of free vibration in a normal mode:

$$T = 2\pi \sqrt{\frac{\sum w_i \delta_i^2}{g \sum f_i \delta_i}}$$

Alternatively, two empirical formulas identical to those of the other codes are given:

$$T = \begin{cases} 0.10\, N & \text{for ductile frames} \\ 0.05\, h/\sqrt{D} & \text{for other buildings} \end{cases}$$

There is a subtle but important difference between the UBC provisions for period calculation and those of the earlier codes, which is discussed in greater detail later.

The Rayleigh quotient formula uses the deflected shape to evaluate the period. Unfortunately, the shape cannot be calculated until the stiffness is known, which requires knowing the design base shear and its distribution, which in turn depend upon the period. It will be necessary to use one of the empirical formulas or some other estimate of the period to get started on any design.

The site-structure resonance factor S is a function of the ratio of the building period to the site period, given by the formulas

$$S = \begin{cases} 1 + \dfrac{T}{T_s} - 0.5\left(\dfrac{T}{T_s}\right)^2 & \text{for } \dfrac{T}{T_s} \leq 1 \\[2em] 1.2 + 0.6\left(\dfrac{T}{T_s}\right) - 0.3\left(\dfrac{T}{T_s}\right)^2 & \text{for } \dfrac{T}{T_s} > 1 \end{cases}$$

but in any case, $S \geq 1$

The relation is shown in Fig. 11. The factor S attains its peak value of 1.5 when the building period and site period coincide, and drops off on either side of resonance to a minimum value of 1 when the building and site periods are widely separated.

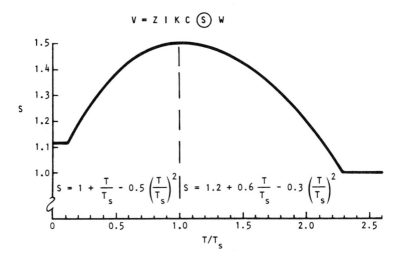

$$V = Z \ I \ K \ C \ \circledS \ W$$

$$S = 1 + \frac{T}{T_s} - 0.5 \left(\frac{T}{T_s}\right)^2 \quad S = 1.2 + 0.6 \frac{T}{T_s} - 0.3 \left(\frac{T}{T_s}\right)^2$$

Figure 11. Site coefficient S, UBC.

To determine the site period T_s requires a geotechnical investigation of the site. UBC Standard No. 23-1 (Ref. 15) gives three methods of determining T_s. One is an analytical method requiring a mathematical analysis of the response of the site, another is an empirical multi-layer method, and the last is an empirical single-layer method. Details of the two empirical methods are specified in Standard 23-1. All three methods require a geotechnical profile of the site showing the physical properties of the soil and rock layers down to bedrock or to a depth of 500 ft., whichever is the lesser, and a knowledge of the shear wave velocities in the different layers. Exception is made for firm sites, where bedrock is shallow and the overlying soil is sufficiently dense, all as defined in Standard 23-1. For such sites, T_s may be taken to be 0.5 sec.

The results of a geotechnical investigation are likely to be a range of site periods rather than a single value. The code stipulates that T_s shall be taken as that value within the determined range that is closest to the building period T. Whatever the results of the geotechnical investigation, the value of T_s used in the formula for S shall be not less than 0.5 sec and not more than 2.5 sec.

The building period T used for determining S must be established by a properly substantiated analysis such as the Rayleigh quotient formula. It may not be empirically determined. Also, for purposes of determining S, the value used for building period T shall not be less than 0.3 sec. This lower bound for T and the upper bound of 2.5 sec for T_s account for showing the cutoff level for S in Fig. 11 as 1.11 instead of 1.00 at the low end of the T/T_s scale.

The UBC provides an alternative to determination of the site period T_s by geotechnical investigaton, which would be to waive the geotechnical investigation and take the site factor to be its maximum value S = 1.5, or if a properly substantiated analysis shows the building period T to exceed 2.5 sec, S may be determined by the appropriate formula with a value T_s = 2.5 sec used for the site period.

In a sense, the product of C and S in the UBC corresponds to the coefficient C in the BOCA, National, and Standard codes and ANSI-72, at least for the zone of greatest seismicity, for which Z = 1. The product C x S is shown as a function of building period in Fig. 12. The lower curve is C x S for the minimum S factor of

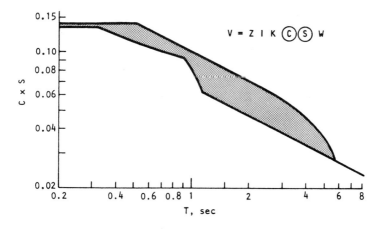

Figure 12. Range of product C x S, UBC.

53

1.0, and the upper curve is for the maximum S factor, which is 1.5 for building periods up to 2.5 sec., decreasing to 1.0 for building periods longer than about 5.7 sec. A cutoff level of 0.14 is stipulated for the product C x S at the short period end of the scale. From Fig. 10, we already had a cutoff level of 0.12 for C, which remains.

Hence, the six factors Z, I, K, C, S, and W give the total base shear, and, as before, the distribution of lateral forces to the various floor levels and to the top of the structure is prescribed. The UBC makes the top force F_t a function of building period rather than of building dimensions:

$$F_t = \begin{cases} 0 & \text{for } T \leq 0.7 \text{ sec.} \\ 0.07 \text{ T V} & \text{for } T > 0.7 \text{ sec.} \end{cases}$$

but in any case, $F_t \leq 0.25$ V

As before, the top force F_t is introduced to accommodate the higher modes by increasing the design shears in the upper stories where the higher modes have the greatest effect. Higher modes are important for tall buildings, which have long periods. If the period is short, higher modes are unlikely to be of much importance in the total response. Note that the UBC makes the maximum value of F_t equal to 25% of the base shear, compared with 15% in the earlier codes. The remainder of the base shear is allocated to the floor levels (including roof level) as before:

$$F_x = \frac{(V - F_t) \, w_x h_x}{\sum w_i \, h_i}$$

The overturning moment for the UBC is calculated as the static effect of the code lateral forces, without a reduction factor.

$$\text{Base OTM} = \sum_{i=1}^{n} F_i h_i + F_t h_n$$

where

F_i, F_t = Lateral force at level i and at the roof
h_i, h_n = Height above base to level i and to the roof

At higher levels a comparable static moment is used. The reduction factor J in the codes discussed earlier was eliminated from the UBC after SEAOC dropped it in 1970, at least in part because of bad results observed in the Caracas, Venezuela, earthquake of 1967. In that earthquake, several tall buildings designed in compliance with seismic building codes much like the American codes had exterior column damage attributed to overturning moment. This is but one of many examples of the modification of seismic building code provisions to reflect experience in real earthquakes.

Drift limits are quantified in the UBC, with the maximum drift allowed in any story specified as .005 times the story height, just as in the National code. The UBC, however, introduces a correction to offset the K factor; for every story

$$\Delta \leq .005 \text{ H} \qquad \text{for K} \geq 1$$
$$\Delta \leq .005 \text{ K H} \quad \text{for K} < 1$$

While it might at first glance seem that more stringent stiffness requirements are being prescribed for more ductile structures, the seeming discrepancy is illusory. The more ductile structure, for which K < 1, enjoys relaxed lateral force requirements. The K factor in the drift limit offsets the reduction of lateral forces, so that the flexibility limits are really the same for all types of structures, independent of the K factor.

As in the codes previously discussed, horizontal torsion is taken into account as a calculated torsion in each story equal to the story shear times the eccentricity of the center of mass above, measured from the center of rigidity of the story. An accidental eccentricity is again prescribed; that is, the torsion is taken as the story shear times either the calculated eccentricity or an accidental eccentricity of 5% of the length of the building, whichever is greater.

Buildings with setbacks in which the plan dimension of the tower in each direction is at least 75% of the corresponding plan dimension of the lower part may be treated as uniform buildings without setbacks if other irregularities do not exist. Otherwise, they must be analyzed with consideration of their special dynamic characteristics.

All buildings designed with K = 0.67 or K = 0.80 are required to have ductile moment-resisting space frames. Buildings more than 160 ft. high must have ductile moment-resisting space frames capable of resisting at least 25% of the seismic force, except that in Zones 1 and 2, they may have concrete shear walls or braced frames if they are designed with either K = 1.00 or K = 1.33. In Zones 2, 3, and 4, all concrete space frames that are essential components of the lateral force-resisting system and all concrete frames located in the perimeter line of vertical support must be ductile moment-resisting frames.

For all buildings in Zones 3 and 4, and for buildings with an importance factor I greater than 1 in Zone 2, all members in braced frames must be designed for 1.25 times the force determined from the UBC seismic analysis. For these buildings, connections must either develop the full capacity of the members or be designed without the 1/3 overstress usually permitted when earthquake forces are present.

The UBC, like the codes discussed earlier, is based on working load concepts. Accordingly, allowable stresses are increased 1/3 in the presence of seismic design forces, a provision that usually affects steel since structural steel is ordinarily designed on a working stress basis. Concrete is designed according to a strength basis. The strength requirements for non-ductile concrete, which is permitted only in Zones 0 and 1, are

$$U = 0.75 \, (1.4 \times \text{dead} + 1.7 \times \text{live} + 1.87 \times \text{earthquake})$$
$$U = \quad (0.9 \times \text{dead} + 1.43 \times \text{earthquake})$$

Both requirements must be met, with the earthquake acting in any direction. The 0.75 factor accomplishes a result comparable to the 1/3 increase in allowable stresses. For ductile moment-resisting space frames of reinforced concrete and for earthquake-resisting concrete shear walls and braced frames, as required for Zones 2, 3, and 4, these requirements are modified to

$$U = 1.40 \text{ (dead load + live load + earthquake)}$$
$$U = 0.90 \times \text{dead load} + 1.40 \times \text{earthquake}$$

The Design Process and Some Notes of Caution

The starting point for designing any building, before considering seismic forces, is the preparation of a preliminary design for dead load, gravity live load, snow, and wind, all of which are treated as static design forces unaffected by the dynamic properties of the structure. The designer can then determine W by estimating the dead load of the building, which will be little affected by the seismic requirements. The type of structure establishes K, the occupancy determines I, and the geographical location Z. The codes give empirical formulas for determining the period T. The designer may make a geotechnical investigation of the site to get the site period T_s, so as to obtain the site-structure resonance factor S; or alternatively, he may elect to waive the geotechnical investigation and take the factor S to be either 1.5 or the maximum value that would pertain for the estimated building period. With this information he can establish the seismic coefficient C, the design base shear V, and the lateral forces F at the various levels, and then design the structural components for the various load combinations that must be considered.

Designers often use some approximate method such as the portal or cantilever method for this preliminary design, proportioning the structural members to meet the stress requirements for the various load conditions. However, stress requirements may or may not be the controlling criteria. The designer must also check drift, to see that it is within code limits, and also torsion, both that due to calculated eccentricity and to accidental torsion. When he has made adjustments so that all requirements are met, he might consider his preliminary design complete. Now — at last — he has all the information at hand to compute a building period according to the UBC Rayleigh formula or some other procedure satisfying the code requirement of properly substantiated technical data. Uncertainties about the extent of cracked concrete sections, the effectiveness of floor slabs, the stiffness contributed by nonstructural components, and the interaction of the structure with its foundation all add to the difficulty of determining the period. The calculated period may differ from that used previously, and therefore may alter the seismic design forces. The designer may then re-cycle the design if he chooses, adjusting member sizes for stress, drift, and torsion.

Upon reaching the point where he considers the preliminary design complete, the designer at last has the information necessary to prepare the input data for a computer analysis using consistent deformations rather than approximate methods. As a matter of fact, most of the programs available for computer-aided design of structures are really analysis programs rather than design programs. They require member properties as input data, and therefore the designer must make some sort of a preliminary design first. Not even then, however, does the computer analysis give the final answer, because the code provides for live load reductions that differ for the different members of a structure according to the type of member, its tributary floor area, and the ratio of dead load to live load. Hence the live load reduction factor may be greater for a main girder than for the floor beams it supports, greater for an interior column than for a corner column, and greater for a lower story column than for a top story column. To accommodate to these factors may require further adjustments in the member sizes, and possibly still another computer analysis of the modified structure.

Different designers may employ countless variations of the process here described, but fundamentally these are the requisite steps, and at the present state of the art, readily available computer programs will not accomplish them without human intervention. The day may come when the computer will do it all, but that day is not here yet.

One of the loops in the process deserves a caveat. The lateral deflections in the UBC Rayleigh quotient formula for period are the *calculated* deflections due to the lateral forces. *Do not calculate the period on the basis of maximum permissible drift!* It might be tempting to take advantage of the maximum flexibility the code allows, for that would lead to the minimum design lateral forces and thus the minimum strength requirements for the structure. True, but there is a flaw in the logic. The drift limit imposed by the code is one-half of one percent of the story height in each story regardless of the seismic zone. The same drift is allowed for all zones, but the lateral forces differ. Thus the Zone 1 structure proportioned for maximum permissible drift would be much more flexible than the Zone 4 structure so proportioned, because the Zone 4 lateral forces are more than five times as great simply on the basis of the Z coefficient. The greater flexibility gives a longer calculated period, thus reducing the seismic coefficient C and further exaggerating the difference in seismic design forces.

It is not at all difficult to calculate the period that would correspond to maximum permissible drift, and it requires no structural analysis at all. For example, consider a building with uniform story weights, a uniform story height of 12 ft. throughout, with the site-structure resonance factor S taken at its maximum value, the structure factor $K = 1$, and the importance factor $I = 1$, proportioned for the maximum permissible drift of .005 times the story height in each story. Such a building four stories high in UBC Zone 4 would have a period of 1.47 sec. In UBC Zone 1 the period would be 5.15 sec. Such a building 50 stories high would have a period of 8.58 sec in UBC Zone 4, and 26.18 sec in UBC Zone 1. A period of 8.58 sec is unreasonably long; a period of 26 sec is absurd. The design base shear for Zone 1, because of the longer period, would be only about 1/9 as great as for Zone 4, instead of the 3/16 that the zone factor would suggest. So,

> Use calculated deflections, not maximum permissible drift, for calculating building period!

One further caveat is related to the site-structure resonance factor S as a function of a period ratio T/T_s, the building period divided by the site period. Recognize that neither the numerator nor the denominator of this fraction can be determined accurately. The numerator is subject to many uncertainties, due to such factors as variations in mechanical properties of materials, dimensional inaccuracies in structural members, deviations from theoretical weights, and imprecise mathematical modeling. The uncertainty in the denominator is even greater. The UBC Standard on determination of T_s speaks of a range of site periods. The mechanical properties of soils simply cannot be determined with accuracy, and rarely are they uniform from one borehole to the next. T, T_s, and S may be computed precisely but still be grossly inaccurate.

APPLIED TECHNOLOGY COUNCIL PROVISIONS

In 1974 the National Science Foundation and the National Bureau of Standards engaged the services of the Applied Technology Council, a research and development subsidiary of the Structural Engineers Association of California, to develop tentative provisions for seismic regulations for buildings. The aim was to write code provisions that could be considered by building authorities all across the nation for possible adoption in their building codes. The Applied Technology Council (ATC) called upon some 85 professionals — engineers, seismologists, geologists, university professors, building officials — to undertake the work in 14 technical committees. The ATC report *Tentative Provisions for the Development of Seismic Regulations for Buildings*, known as ATC 3-06, was published jointly by ATC, NSF, and NBS in 1978. Its provisions are now being tested, reviewed, and evaluated. They depart from earlier code provisions in some respects and are similar in others. The key provisions will be presented here, and compared with earlier codes where appropriate.

In the first place, there are fundamental philosophical differences between ATC 3-06 and its predecessors. Earlier codes provided equivalent static lateral forces predicated upon the regional seismicity, building type, period, and, in the case of the UBC, the occupancy and a soil-structure interaction factor. Seismic zones were based on the estimated extreme ground motion of the region in the historical past, without regard to the frequency of occurrence of earthquakes. Other factors such as building type and occupancy either increased or decreased the design lateral forces. Thus a building in Buffalo or Paducah —Zone 3 — would be subject to the same code requirements as a building in Seattle or San Diego — also Zone 3 — even though their annual exposure to earthquake risk would be substantially less. Moreover, the same design procedures would be required for a warehouse or factory building as for a fire station or hospital, although the importance factor in the UBC would affect the magnitude of the design lateral forces.

The Applied Technology Council expresses seismicity in terms of zone maps that take both intensity of ground motion and frequency of occurrence into account, and it imposes different restrictions as to permissible structural systems or permissible building design procedures according to the degree of loss that society would incur from failure of the building. Their purpose is to establish nationwide a more or less uniform annual earthquake risk. Clearly, the achievement of a uniform annual earthquake risk is not actually an attainable goal — indeed, it is hardly even definable. The probability of building failure that society would consider an acceptable risk for a single isolated building might be quite different from the acceptable risk for a group of a hundred similar buildings in a city. Nonetheless, despite the recognized impossibility of the task, the ATC has attempted to establish some degree of national uniformity in annual earthquake risk.

The ATC 3-06 provisions contain several significant departures from earlier seismic design procedures. Regional seismicity is characterized by more realistic ground motion indices that account not only for extreme motion at the source, but also for attenuation with distance. The single zone coefficient of earlier codes has been replaced by two indices related to effective peak acceleration and effective peak velocity, plus a seismicity index. Buildings are classified by use into seismic hazard exposure groups. The seismicity index and seismic hazard exposure group determine a seismic performance category for the building that regulates the design and analysis requirements. In general, the analysis and design procedures in this code are based upon a limit state characterized by significant yield rather than upon working load as in earlier codes. In view of these departures, no simple yet meaningful comparison can be made between the new ATC 3-06 provisions and those of earlier codes to establish that one set is more severe or more lenient than others.

Seismic Performance Categories

To determine the appropriate seismic performance category, ATC 3-06 assigns buildings to different seismic hazard exposure groups according to their uses. Together, the seismic hazard exposure group and the seismicity index determine the seismic performance category for the building. The ATC 3-06 provisions leave it to the building authorities to establish exact definitions of the seismic hazard exposure groups, but they provide guidelines that we summarize here:

Seismic Hazard Exposure Groups

Group	Description
III	Buildings used for functions that are essential to post-earthquake recovery, such as fire stations, police stations, power stations or other necessary utilities, medical facilities offering surgery and emergency treatment, and emergency preparedness centers.
II	Buildings housing large numbers of occupants or occupants of restricted or impaired mobility, such as public assembly buildings, large open-air stands, large retail stores or shopping centers, and buildings more than 4 stories high used for offices, hotels, apartments, or factories; hospital buildings not included in Group III; buildings used for day-care centers, schools, or colleges; and buildings used for hazardous purposes involving the storage or use of flammable or toxic liquids.
I	Buildings not in Group II or Group III.

The input ground motion is represented by two indices instead of one because different frequencies of input motion have different effects on buildings with different dynamic properties. Short period structures are sensitive mainly to short period components of ground motion, whereas longer period structures are relatively immune to short period motion and sensitive to the longer period components.

The coefficient A_a, which characterizes the short period components of ground motion, is determined from the zone map shown in Fig. 13. It is equal to the effective peak acceleration divided by the acceleration of gravity. Effective peak acceleration is not the maximum ground acceleration, but is related to the response spectral acceleration averaged over the short period range of 0.1 to 0.5 sec and divided by a normalizing factor. It is usually less than the maximum instantaneous ground acceleration. The values of the coefficient for the seven seismic zones shown in Fig. 13 are

Seismic Coefficient, A_a

Zone	A_a
7	0.40
6	0.30
5	0.20
4	0.15
3	0.10
2	0.05
1	0.05

Fig. 14 shows the zone map for determining the coefficient A_v and the seismicity index. A_v characterizes the longer period components of ground motion; it is a dimensionless coefficient related to the effective peak velocity, the response spectral velocity at a period of about 1.0 second divided by a normalizing factor. Unlike the effective peak acceleration, which is usually less than the peak ground acceleration, the effective peak velocity may exceed the maximum instantaneous ground velocity. The values of the coefficient A_v and the seismicity index for the seven zones shown in Fig. 14 are

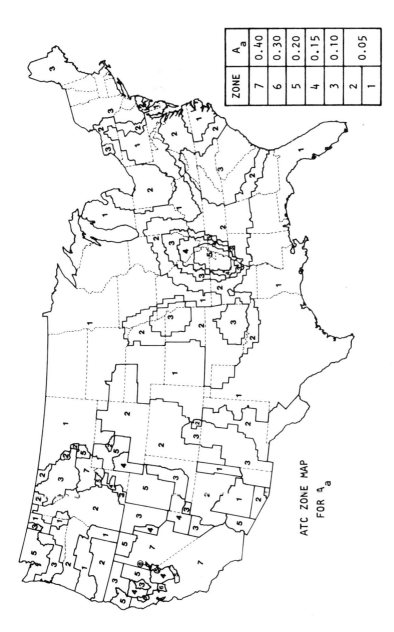

ZONE	A_a
7	0.40
6	0.30
5	0.20
4	0.15
3	0.10
2	0.05
1	

ATC ZONE MAP
FOR A_a

Figure 13. Applied Technology Council zone map for seismic coefficient A_a.

65

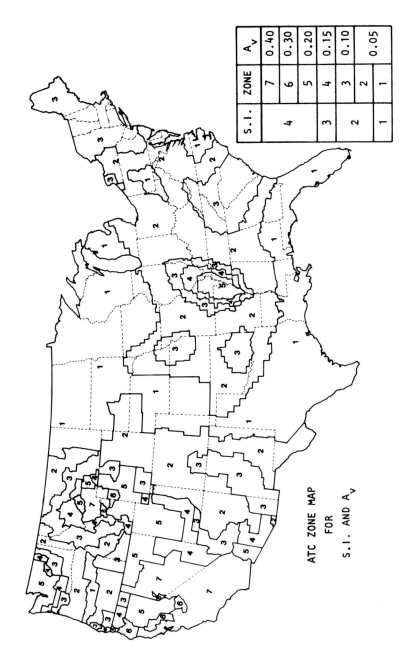

S.I.	ZONE	A_v
4	7	0.40
	6	0.30
	5	0.20
3	4	0.15
2	3	0.10
	2	0.05
1	1	

ATC ZONE MAP
FOR
S.I. AND A_v

Figure 14. ATC 3-06 zone map for seismic coefficient A_v and Seismicity Index.

Seismic Coefficient A_v and Seismicity Index

Zone	A_v	Seismicity Index
7	0.40	4
6	0.30	4
5	0.20	4
4	0.15	3
3	0.10	2
2	0.05	2
1	0.05	1

Unlike the seismic zones in earlier codes, which were based on historical extreme motion, the ATC 3-06 zones are based on probability of occurrence. The zone boundaries for A_a and A_v follow approximately the contours along which there is a 90% probability that the appropriate values of effective peak acceleration and effective peak velocity will not be exceeded in any 50-year period. The zone boundaries have irregular shapes because they follow political boundaries, a practical concession intended to facilitate code enactment and enforcement.

ATC 3-06 employs Seismic Performance Categories for buildings to establish what types of structural systems may or may not be used, and what procedures are required for structural analysis and design. The Seismic Performance Category is assigned according to the Seismic Hazard Exposure group and the Seismicity Index for the location. We first present the ATC 3-06 definitions of structural framing systems and then the performance categories.

ATC 3-06 defines four types of structural framing systems acceptable for buildings:

(1) Bearing wall system, in which bearing walls provide all or most of the support for gravity loads and shear walls or braced frames provide the seismic resistance

67

(2) Building frame system, in which a space frame supports gravity loads and shear walls or braced frames provide the seismic resistance

(3) Moment-resisting frame system, in which a space frame supports gravity loads and an ordinary or special moment frame is capable of resisting the total prescribed seismic forces

(4) Dual system, in which a space frame supports gravity loads, a special moment frame is capable of resisting at least a quarter of the prescribed seismic forces, and the special moment frame and shear walls or braced frames together resist the total seismic force, sharing the load in proportion to their relative rigidities

Not included in these four structural systems are inverted pendulum structures, in which both gravity and seismic loads are supported by a framing system that acts essentially as one or more isolated vertical cantilevers. ATC 3-06 also gives design information for these systems.

There are four Seismic Performance Categories:

Category A buildings may be designed to use any of the four defined types of framing systems. They need not be analyzed for seismic forces exerted on the building as a whole. Minimum requirements for ties, continuity, and anchorage of walls and nonstructural components are stipulated.

Category B buildings may also be designed to use any of the four types of framing systems. They must be analyzed, however, by at least an equivalent lateral force procedure, to be described in more detail below. This procedure is roughly the same as that used for the codes discussed earlier, although the requirements differ. There are also material and construction limitations that are more restrictive than those for Category A.

Category C imposes additional material and construction limitations. Category C buildings more than 160 ft. tall must be either moment-resisting frame structural systems (type 3 above) with special moment frames; dual structural systems (type 4); or building frame systems (type 2) with structural steel or cast-in-place concrete braced frames or shear walls resisting lateral forces, but subject to limitations on the design force resisted by the walls or frames in any one plane. Building frame systems (type 2) are limited to a height of 240 ft. Regular buildings in Category C must be analyzed at least by an equivalent lateral force procedure. If irregularities exist, a more rigorous method of analysis is required, ordinarily a modal analysis. Vertical earthquake effects must be considered for horizontal cantilevers and for horizontal prestressed components.

For *Category D* buildings there are the same requirements as for Category C with regard to structural types but with more restrictive height limits. The height above which special moment frame systems, dual systems, or restricted building frame systems are required is reduced from 160 ft to 100 ft., and the height limit for the restricted building frame systems is reduced from 240 ft. to 160 ft. Analysis requirements are the same as for Category C.

The combination of the requirements for Category C buildings and analysis by the equivalent lateral force procedure is roughly the same as for current practice in Zone 4 under the UBC for buildings not classified as essential facilities.

The combination of Seismic Hazard Exposure Group for the building and Seismicity Index for the location determines which Seismic Performance Category will pertain to the building, according to the following:

Seismic Performance Category

Seismicity Index	Seismic Hazard Exposure Group		
	III	II	I
4	D	C	C
3	C	C	B
2	B	B	B
1	A	A	A

Equivalent Lateral Force Procedure

The equivalent lateral force procedure, the minimum level of analysis acceptable for all buildings in Seismic Performance

Category B and for regular buildings in Categories C and D, is similar in principle to the procedures of earlier codes. The effects of the earthquake are represented by static lateral forces.

The design base shear is given by the formula

$$V = C_s W$$

in which C_s is the seismic design coefficient and W is the total gravity load. W specifically includes partitions and permanent equipment, plus effective snow load. Effective snow load is taken to be 70% of the design gravity snow load, but the fraction may be reduced to a minimum of 20% of design snow load if approved by the cognizant regulatory agency. Freshly fallen snow has little influence on lateral seismic forces, but ice accumulation on the roof would have a great influence. For storage and warehouse occupancies, W includes 25% of the floor live load as well.

C_s is the lesser value calculated value from the two formulas

$$C_s = \frac{1.2 \, A_v \, S}{R \, T^{2/3}}$$

$$C_s = \frac{2.5 \, A_a}{R} \quad \text{(with an exception noted later)}$$

where

A_a = Effective peak acceleration coefficient
A_v = Effective peak velocity coefficient
S = Site coefficient, related to the soil profile
T = Fundamental period of the building, sec
R = Response modification factor for the structural system

A_a and A_v are determined from the zone maps, Figs. 13 and 14.

As in the codes discussed earlier, the first formula provides a seismic design coefficient that decreases with increasing period except that the exponent of the period in the denominator is 2/3 here, compared with 1/2 in UBC and 1/3 in the other codes. The

70

second formula essentially provides a cutoff level at the short period end of the scale.

From the exact definitions in the ATC 3-06 report of the three types of soil profile from which to determine the site coefficient S, Table 4 presents the following approximate descriptions of the soil profile:

Table 4

Site Coefficient, S

Soil profile	S
Rock or shallow stiff soil over rock	1.0
Deep stiff soil over rock	1.2
Soft soil	1.5

The exception mentioned above for the second formula for C_s occurs if $S = 1.5$ and $A_a \geq 0.30$. In such a case, the seismic coefficient is $C_s = 2.0\, A_a/R$ instead of $2.5\, A_a/R$.

In the response modification factor R, the ATC 3-06 provisions recognize that structures generally have a reserve capacity to absorb energy beyond the elastic strain energy at significant yield. The R factor essentially represents the ratio of the extreme forces that would develop in linear elastic response to the specified ground motion, to the static forces that would cause significant yield. It takes into account the ductility of the structural system. The reducton in design base shear is possible because the capability of a ductile structure to deform inelastically provides an ability to absorb and dissipate energy fed into the structure by the ground motion.

The ATC 3-06 report gives a table of R values for the different classes of structural systems, depending on the kind of subsystem to be used to resist seismic forces. The ranges of values of R are given in Table 5.

Table 5

Response Modification Factor, R

Structural System	Range of R
Bearing wall system (type 1)	From 1-1/4 for unreinforced masonry walls to 6-1/2 for light framed walls with shear panels.
Building frame system (type 2)	From 1-1/2 for unreinforced masonry shear walls to 7 for light framed walls with shear panels.
Moment-resisting frame system (type 3)	From 2 for ordinary moment frames of reinforced concrete to 8 for special moment frames of steel.
Dual system (type 4)	From 6 for braced frames to 8 for reinforced concrete shear walls.
Inverted pendulum structure	From 1-1/4 for ordinary moment frames of steel to 2-1/2 for special moment frames of either steel or reinforced concrete.

A considerable amount of professional judgment has gone into the establishment of the R factors, with theory, experiment, and the behavior of different materials and structural systems in past earthquakes all taken into account.

The fundamental period, T, of the structure may be taken as the approximate period T_a given by whichever of the following three empirical formulas applies:

For moment-resisting frame structures with the frames not constrained by more rigid components,

$$T_a = \begin{cases} 0.035 \ h^{3/4} & \text{for steel frames} \\ 0.025 \ h^{3/4} & \text{for concrete frames} \end{cases}$$

For all other buildings,

$$T_a = \frac{0.05 \ h}{\sqrt{L}}$$

where
   h = Height of highest level of the building above the base, ft.
   L = Overall length of the building at the base in the direction under consideration, ft.

Alternatively, the period T may be computed by established methods of mechanics based on the elastic properties of the seismic resisting system. The Rayleigh quotient formula for period given in the Uniform Building Code would be one such method of mechanics. However, the period T used in the formula for base shear may not in any case exceed the approximate period T_a calculated from the foregoing empirical formulas by more than 20%. This requirement provides a safeguard against inadvertent misapplication of the Rayleigh formula as discussed earlier for the Uniform Building Code.

The base shear calculated from the foregoing formulas for the equivalent lateral force procedure may be reduced to account for the effects of the interaction of the structure with the soil on which it is based. The reduction is complex; indeed, a complete chapter in the ATC 3-06 report is devoted to it. In brief, the reduction recognizes that if the foundation were not rigid, then the motion that would occur at the foundation of the building would differ from the free-field ground motion, that is, the ground motion that would occur at the foundation location if the structure were absent. The effects of the interaction of the

73

structure with the underlying soil are to lengthen the period of the building from the fixed-base condition due to both lateral motion and rocking of the foundation, and to modify the effective damping — usually to increase it — due to hysteretic action in the supporting medium and to radiation of energy into the surrounding earth.

The amount that the base shear may be reduced because of soil-structure interaction depends upon the calculated periods of the structure with fixed base and with foundation compliance, and upon the fraction of critical damping for the structure-foundation system. These in turn are functions of the stiffness and height of the building and the lateral stiffness and rocking stiffness of the foundation. Formulas are given in the ATC 3-06 report. In any event, the reduced base shear must be at least 70% of the base shear calculated without consideration of soil-structure interaction.

The base shear V is distributed as lateral forces F_x at the different levels of the building according to the formula

$$F_x = C_{vx} V$$

where

$$C_{vx} = \frac{w_x h_x^k}{\sum w_i h_i^k}$$

and

$w_x, w_i =$ Weights at levels x and i
$h_x, h_i =$ Heights of levels x and i
        above the base

$$k = \begin{cases} 1 & \text{for } T \leq 0.5 \text{ sec} \\ 2 & \text{for } T \geq 2.5 \text{ sec} \\ 0.75 + T/2 & \text{for } 0.5 < T < 2.5 \text{ sec} \end{cases}$$

The distribution is shown in Fig. 15. It is linear for short period structures, for which higher modes are of little consequence, and parabolic for long period structures, for which higher modes are more significant. In the parabolic distribution, a greater portion of the base shear is assigned to the higher levels, thereby increasing the story shears at the upper levels and increasing the overturning moments throughout. For intermediate period structures there is a smooth transition.

Horizontal torsion is to be considered in each story, calculated as the torsion due to the lateral forces F at the levels above, each acting with an eccentricity determined by the location of the center of mass at that level, plus an additional eccentricity (in either direction) equal to 5% of the plan dimension of the building normal to the direction of the seismic forces being considered. The codes discussed earlier used *either* the calculated eccentricity *or* an accidental eccentricity equal to 5% of the *long* plan dimension of the building, whichever was greater. ATC 3-06 uses the calculated eccentricity *plus or minus* an accidental eccentricity of 5% of the plan dimension of the building in the direction *normal to the seismic forces* being considered.

FORCE DISTRIBUTION

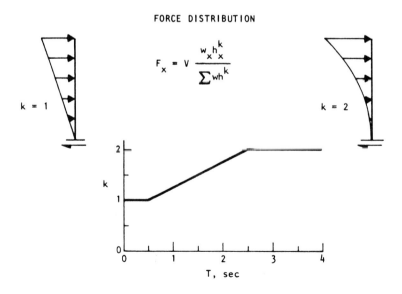

$$F_x = V \frac{w_x h_x^k}{\sum w h^k}$$

Figure 15. Lateral force distribution exponent k, ATC 3-06.

75

The overturning moment M_x at level x is computed from the static effects of the lateral forces F_i above that level, modified by a reduction factor κ, thus:

$$M_x = \kappa \sum_{i=x}^{n} F_i (h_i - h_x)$$

where

F_i = Lateral force at level i
h_i, h_x = Heights of levels i, x above the base

$$\kappa = \begin{cases} 1 \text{ for the top ten stories} \\ 0.8 \text{ for the 20th story from the top and below} \\ \text{is determined by interpolation between} \\ \text{these values for stories between these levels} \end{cases}$$

Except for inverted pendulum structures, the foundation design overturning moment M_f at the foundation-soil interface is determined by the same formula, with κ taken to be 0.75 for all building heights, with the further restriction that the resultant of the lateral seismic forces and vertical gravity loads must fall within the middle one-half of the base of the components resisting the overturning.

The reduction factor κ compensates for phase differences between the lateral forces at the various levels. The design lateral forces are intended conservatively to envelop the story shears, but they would not reach their extreme values at all levels in the same direction simultaneously. Thus the overturning moment calculated as the static effect of the design lateral forces would be overconservative. The BOCA, National, and Standard codes and ANSI-72 provided more of a reduction than ATC 3-06; but the exceptionally large overturning forces that occurred in the 1967 Caracas earthquake, as inferred from the damage to columns in several tall buildings, led UBC to eliminate the reduction altogether. The ATC 3-06 provisions are a compromise.

ATC 3-06 departs from the drift limits of earlier codes, largely because the procedures are based on limit state instead of working load design. First, the lateral displacement δ_{xe} at each level is calculated as the elastic displacement of the seismic resisting system with fixed base due to the design lateral forces F_x. Alternatively, the deflections may be calculated for the seismic resisting system using the forces corresponding to the fundamental period T, calculated according to a method of mechanics such as the Rayleigh quotient formula without regard to the limit $T \leq 1.20\ T_a$. This alternative requires, of course, that the same model of the seismic resisting system be used for calculating deflections that is used for calculating the period T. The deflection calculations must include rotation of joints, both flexural and axial deformations of frame members, and both shear and flexural deformations of shear walls and braced frames.

The elastic displacements δ_{xe} are multiplied by a deflection amplification factor C_d to get the displacements δ_x, thus:

$$\delta_x = C_d\ \delta_{xe}$$

The amplification factor is used because the denominator in the base shear formula includes a response modification factor R to account for the capability of the structure to dissipate energy in inelastic deformation. The multiplier C_d partially offsets the divisor R, so that the reduced strength requirement does not carry with it an equal reduced stiffness requirement. The two coefficients R and C_d do not differ greatly, but they are not identical.

The ATC 3-06 report gives C_d values for the different classes of structural systems depending on the type of seismic resisting system employed, in the same table as the R values. The ranges for C_d are given in Table 6.

The story drift Δ is the difference between the displacements δ_x at the floor levels above and below the story. The maximum story drift in any story is limited to 0.01 times the story height for buildings in Seismic Hazard Exposure (SHE) Group III, and to 0.015 times the story height for buildings in SHE Groups II and I. For buildings of three or fewer stories in SHE Group I, the limit may be increased to 0.02 times the story height if there are no brittle-type finishes in the building.

Table 6

Deflection Amplification Factor, C_d

Structural System	Range of C_d
Bearing wall system (type 1)	From 1-1/4 for unreinforced masonry walls to 4 for either light framed walls with shear panels or reinforced concrete shear walls.
Building frame system (type 2)	From 1-1/2 for unreinforced masonry walls to 5 for reinforced concrete shear walls.
Moment-resisting frame system (type 3)	From 2 for ordinary moment frames of reinforced concrete to 6 for special frames of reinforced concrete.
Dual system (type 4)	From 5 for braced frames to 6-1/2 for reinforced concrete shear walls.
Inverted pendulum structure	From 1-1/4 for ordinary moment frames of steel to 2-1/2 for special moment frames of either steel or reinforced concrete.

P-Δ effects result from the moment due to the gravity loads above being displaced laterally. Usually they are not important, but they do lead to an increase in member forces and story drifts, and may be significant under some circumstances. ATC 3-06 introduces a stability coefficient θ, determined by the formula

$$\theta_i = \frac{P_i \, \Delta_i}{V_i h_{si}}$$

where

θ_i = Stability coefficient for story i
P_i = Total gravity load above story i
Δ_i = Design story drift in story i
V_i = Total shear in story i
h_{si} = Height of story i

If $\theta_i \leq 0.10$ for every story, then the P-Δ effects may be ignored. If $\theta_i > 0.10$ for any story, then the story drifts must be increased to account for the P-Δ effects, and the effects of this drift increment on the story shears and moments must be taken into account. ATC 3-06 stipulates only that the P-Δ effects must be determined by rational analysis. A commentary in ATC 3-06 outlines an acceptable analysis.

The building must be designed for the following combinations of dead load, live load, snow load, and earthquake:

(1) 1.2 x dead + 1.0 x live + 1.0 x snow $\pm$ 1.0 x earthquake

and *either*

(2a) 0.8 x dead $\pm$ 1.0 x earthquake

or, for splices in steel columns that use partial penetration welds as well as for unreinforced masonry and other brittle materials, systems, and connections,

(2b) 0.5 x dead $\pm$ 1.0 x earthquake.

The design must be adequate both for loading 1 and for loading 2a or 2b, as applicable.

Orthogonal effects must also be considered. They may be satisfied by combining 100% of the effects of earthquake motion in one principal direction with 30% of the effects of motion in the perpendicular direction, using the combination that gives the most severe effect. This requirement may often influence the design of corner columns, for they more than other members feel the effect of overturning moment in both principal directions.

For buildings in Seismic Performance Categories C and D, the vertical component of earthquake motion must be considered in the design of horizontal cantilever members and horizontal prestressed members. For horizontal cantilevers, this requirement may be satisfied by designing for a net upward force of 20% of the dead load. For other prestressed horizontal components, design for the last of the three load combinations given above is adequate to account for the vertical component of earthquake motion.

The foregoing equivalent lateral force procedure is suitable for the design of most buildings. Buildings in seismic performance categories C and D that are classified as irregular, such as buildings having abrupt changes in their stiffness or inertia properties, must be analyzed with special consideration for their dynamic characteristics. In most cases, this requirement may be met by the modal analysis procedure given in the ATC 3-06 report.

In brief, the modal analysis procedure is a classical normal modes analysis with the structure treated as a lumped-mass system, with one mass at each floor level. For each principal direction, the analysis must include either three modes of vibration or all modes with periods longer than 0.4 sec., whichever is the greater number. An exception is made for structures with three or fewer stories, for which the number of modes in each direction must be equal to the number of stories.

Modal base shears are computed using modal effective weights and modal seismic coefficients determined by a procedure comparable to that of the equivalent lateral force method. The modal base shears are allocated to the various levels as lateral forces determined according to the mode shape, and modal story drifts are computed using deflection amplification factors as for the equivalent lateral force procedure. Design base shears, story shears, moments, and drifts are computed as the square root of the sum of the squares of the corresponding modal quantities. A few of the requirements are modified somewhat from those of the equivalent lateral force procedure.

ATC 3-06 modifies the standard AISC and ACI specifications and procedures to make them compatible with its limit-state force levels. For steel design it provides the following capacity reduction factors ϕ:

For members and for connections that
develop the strength of the members
$\phi = 0.90$

For connections that do not develop
the strength of the members
$\phi = 0.67$

For partial penetration welds in columns
when subjected to tension stresses
$\phi = 0.80$

Then, instead of the normal 1/3 increase in allowable stresses, ATC 3-06 provides that the strength of steel members resisting seismic forces, either alone or in combination with other loads, shall be determined by using 1.7 times the allowable stresses in the pertinent AISC specifications. For shear, however, ATC 3-06 uses 0.32 F_y for the allowable shear stress instead of the 0.40 F_y stipulated by AISC; that is, the shear strength is determined from a stress of 1.7 times 0.32 F_y instead of 1.7 times 0.40 F_y.

For reinforced concrete, ATC 3-06 modifies the ACI capacity reduction factors as follows:

For connections of precast components
$\phi = 0.50$

For axial compression alone or axial compression combined with bending on any member where axial stress exceeds $0.10f_c'$ and the axial stress due to seismic forces exceeds $0.05f_c'$ and special lateral reinforcement is not provided
$\phi = 0.50$

For shear strength of components of normal weight concrete in Seismic Performance Category C and D buildings:
If strength is governed by flexure $\phi = 0.85$
If strength is governed by shear $\phi = 0.60$

For shear strength of lightweight concrete $\phi = 80\%$ of the value for normal weight concrete.

81

CODE COMPARISON

It is difficult to make a general comparison of the results of these different codes because of their different approaches. We can, however, take specific cases and compare the code requirements for base shear capacity, reduced to a common working load level.

Consider a multi-story office building with a moment-resisting steel framing system meeting all of the requirements for a ductile moment-resisting frame according to the four current codes, and also meeting the ATC 3-06 requirements for a special moment frame. Assume that the site has deep, stiff soil overlying rock, and that there will be no geotechnical investigation of the site.

Comparative Requirements in Maximum Seismic Zone

First, take the location of the building to be somewhere in the maximum seismic zone for each of the codes, which is Zone 3 under BOCA, National, Standard, and ANSI-72, Zone 4 under UBC, and Zone 7 under ATC 3-06. Parts of California and Nevada would fall in these zones.

The BOCA, National, and Standard codes and ANSI-72 all stipulate a base shear

$$V = ZKCW$$

where

$Z = 1$ for Zone 3
$K = 0.67$ for a ductile moment-resisting frame
$C = .05 / \sqrt[3]{T}$, with the limit $C \leq 0.10$

These codes all contain a seismic overload factor of $4/3$, either by allowing a one-third overstress or by providing a load reduction factor of $3/4$ if earthquake forces are present. Thus the base shear capacity requirement at working load level is

$$V_w = 1 \times 0.67 \times (.05/\sqrt[3]{T}) \times (3/4) \times W$$

$$\approx .025 \ W/\sqrt[3]{T}$$

subject to the limit

$$V_w \leq 1 \times 0.67 \times 0.10 \times (3/4) \times W \approx 0.050 \ W$$

The UBC uses a base shear

$$V = ZIKCSW$$

where

$Z = 1$ for Zone 4
$I = 1$ for an office building
$K = 0.67$ for a ductile moment-resisting frame
$C = 1/(15\sqrt{T})$, subject to the limit $C \leq 0.12$
$S = 1.5$ in the absence of a geotechnical investigation
$CS \leq 0.14$

If the building period were longer than 2.5 sec, then S could be reduced, but we will presume this is not the case. The UBC also provides a one-third increase in allowable stresses for steel. Thus the base shear capacity, reduced to working load level, is

$$V_w = 1 \times 1 \times 0.67 \times 1.5 \times (3/4) \times W / (15\sqrt{T})$$

$$\approx 0.050 \ W/\sqrt{T}$$

subject to the limit

$$V_w \leq 1 \times 1 \times 0.67 \times 0.14 \times (3/4) \times W \approx 0.07 \ W$$

The ATC 3-06 provisions call for a base shear

$$V = 1.2 \ A_v \ S \ W \ / \ (R \ T^{2/3})$$

but limited by

$$V \leq 2.5 \ A_a \ W \ / \ R$$

where

$A_a = 0.40$ for Zone 7
$A_v = 0.40$ for Zone 7
$S \ = 1.2$ for deep stiff soil over rock
$R \ = 8$ for a steel moment-resisting frame system meeting the criteria for a special moment frame

The seismicity index is 4 in Zone 7 of the A_v zone map, which would place this building in Seismic Performance Category C. The special moment frame system chosen for this example and an equivalent lateral force method of analysis would satisfy the requirements for buildings in Category C.

ATC 3-06 calls for steel member strength to be evaluated by plastic design procedures, using 1.7 times the bending stress allowed for conventional elastic design. It also imposes a capacity reduction factor $\phi = 0.90$ for steel members and connections that develop the strength of the members. Combining these and taking the ratio of plastic moment to yield moment to be $M_p / M_y = 1.14$, which is about average for rolled steel beams, we get the base shear capacity requirement, reduced to working load level, to be

$$V_w = \frac{1.2 \times 0.40 \times 1.2 \times W}{0.9 \times 1.7 \times 1.14 \times 8 \times T^{2/3}}$$

$$= \frac{0.0413 \ W}{T^{2/3}}$$

subject to the limit

84

$$V_w \le \frac{2.5 \times 0.40 \times W}{0.9 \times 1.7 \times 1.14 \times 8} \approx 0.072 \, W$$

Fig. 16 shows the comparative requirements. For buildings with short periods, the ATC 3-06 capacity requirement is the greatest, just slightly more than that of UBC, and BOCA/ National/Standard/ANSI-72 is somewhat lower. For buildings with periods longer than about half a second, the UBC requirement exceeds that of ATC 3-06.

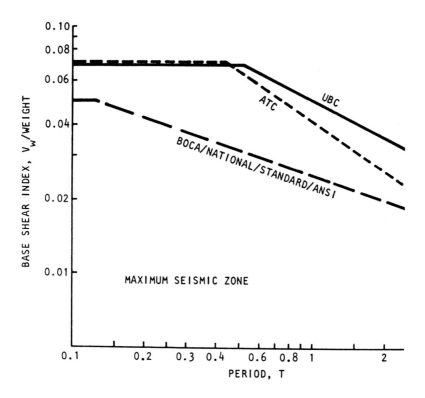

Figure 16. Comparative capacity requirements, example building in maximum seismic zone.

Comparative Requirements in Buffalo

Now consider another multi-story office building utilizing the same structural system, with the same site conditions, but located instead in the city of Buffalo, New York. Buffalo is in the zone of highest seismicity in BOCA/National/Standard/ANSI-72, in which there are three zones numbered 1 to 3; the second highest in UBC, which has five zones numbered 0 to 4; and about mid-range among all the zones in ATC 3-06, which has seven zones numbered 1 to 7. By coincidence the zone number for Buffalo is the same, Zone 3, for all of the codes.

All of the foregoing calculations would be the same except for the coefficients related to the seismicity of the region. Those coefficients are

$$Z = 1 \text{ for BOCA/National/Standard/ANSI-72 Zone 3}$$
$$Z = 3/4 \text{ for UBC Zone 3}$$
$$A_a = 0.10 \text{ for ATC 3-06 Zone 3}$$
$$A_v = 0.10 \text{ for ATC 3-06 Zone 3}$$

The ATC 3-06 seismicity index is 2 for A_v Zone 3, which would place this building in Seismic Performance Category B. The lower category would relax some of the restrictions as to permissible building systems, but it would have no effect on this particular example.

The base shear capacity requirement would be exactly the same as before for the BOCA, National, and Standard codes and ANSI-72; 3/4 as great as before for UBC; and only 1/4 as great as before for the ATC 3-06 provisions. Figure 17 shows the comparative results. Here the UBC capacity requirement is the greatest of the three for all periods in the range shown, with BOCA/National/Standard/ANSI-72 not far behind. The ATC 3-06 capacity requirement is far below the others over the entire range of periods.

This is something of a gross comparison that is not truly indicative of the comparative member design requirements, for ATC 3-06 requires different load combinations than the other codes, and it requires a consideration of orthogonal effects of the earthquake.

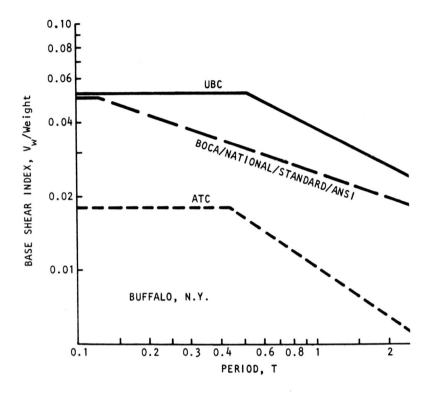

Figure 17. Comparative capacity requirements, same building but located in Buffalo, N.Y.

EXPERIENCE IN PAST EARTHQUAKES

"From Mallet's time to the present day," wrote Charles Richter, "reports have confirmed, in melancholy repetition, the obvious fact that defective construction will not withstand an earthquake" (Ref. 16). And, to be sure, earthquake reports repeatedly demonstrate the obvious. Nevertheless, field studies have led to code improvements, and they have provided indispensable lessons for the structural engineer. Let us consider a few.

The merits of ductility were vividly illustrated by the behavior of the columns in the ground story of Olive View Hospital in the earthquake of 1971 in San Fernando, California (Ref. 17). Figure 18 shows one of the corner columns, a tied column, completely shattered by the distortion imposed on it. Figure 19 shows a spiral column in the same building, with virtually identical lateral distortion, still able to support weight from the building above it.

The vast difference in behavior between these two columns is primarily due to the difference between shear and flexural failure of reinforced concrete. In a tied column such as that of Fig. 18, the ties yield or break in tension, and concurrently the concrete fractures in shear. Failure is brittle, and the column abruptly loses its capacity to bear vertical load. In contrast, in a spiral column such as that of Fig. 19, the outside shell disintegrates, but the spiral steel confines the core concrete and inhibits brittle shear fracture. The core incurs flexural failure due to the lateral displacement, with the concrete crushed in vertical compression on the concave side and the compression relieved or reversed on the convex side. Deterioration is gradual and ductile in nature, and much of the load-carrying capacity of the column persists far beyond the distortion at which deterioration commences.

According to the principle of relative rigidity, one of the principles of earthquake-resistant construction postulated by Dr. Naito nearly sixty years ago, the resisting components of a structure will share the lateral forces of an earthquake in proportion to their relative stiffnesses. In the classroom building of Figs. 20 and 21 in Chimbote, Peru, damaged in the 1972 earthquake, one side of the building had low masonry walls and

Figure 18. Tied reinforced concrete column, Olive View Hospital.

Figure 19. Spiral reinforced concrete column, Olive View Hospital.

90

Figure 20. Columns adjoined by tall windows, Colegio Regional, Chimbote, Peru.

Figure 21. Columns adjoined by short windows, Colegio Regional, Chimbote.

tall windows between the columns, while the opposite side had high masonry walls and short windows (Ref. 17). For lateral displacement in the plane of the wall, each column was essentially fixed at the beam above and at the masonry wall below the window. The short columns were much stiffer than the long columns, and therefore took a greater share of the earthquake load. Both short and long columns failed, the long ones in flexure and the short ones in shear. The long columns retained part of their capacity to support vertical loads; the shorts ones lost all of theirs.

In the context of earthquake-resistant construction, "nonstructural elements" is a misnomer, for actually, any element in a building, whether an essential component of the structural system or not, affects the behavior of the building unless it is isolated so that it remains completely undeformed. Nonstructural masonry walls and partitions are usually built in contact with the adjoining structural frame, and, even though they are not designed to do so, they provide strength and stiffness up to their capacity, thereby influencing the response of the building. This may be either a help or a hindrance. The nonstructural masonry walls of the Chimbote school building (Figs. 20 and 21) made the effective lengths of the columns on one side of the building much shorter than on the other side, giving them greater rigidity. This contributed to their destruction, but it also helped limit the total displacement of the building.

The four-story Banco Nor Peru building in Chimbote (Figs. 22 and 23) also exemplifies the principle of relative rigidity (Ref. 18). In the ground story the front transverse wall was almost entirely glass, and the rear transverse wall was offset from the upper part of the building and structurally separated from the main building. Thus the columns were virtually the only structural components available to resist transverse motion in the ground story.

Figure 23 shows some of the interior columns. Those near the front of the building were laterally unrestrained for the full story height, whereas those toward the rear were supported at mid-height by a mezzanine floor, and the column at the back was further restrained in the transverse direction by a low masonry wall alongside the stairway. The columns toward the front,

Figure 22. Banco Nor Peru, Chimbote.

Figure 23. Column damage, Banco Nor Peru, Chimbote.

having the greatest unrestrained length, were the most flexible and showed the least damage. Those that were restrained at mid-height by the mezzanine floor were more rigid because of their shorter effective length, and they encountered greater damage in shear and flexure. The rear column, being further restrained by the masonry wall adjacent to the stairway, had a very short effective length, and was thus relatively quite stiff. It was completely sheared.

The Banco Nor Peru building illustrates the phenomenon of the ground story that is considered soft. Like many commercial buildings, it had an open ground story with no interior walls and only a few lightweight counters and interior partitions. Masonry exterior walls provided ground story stiffness in the longitudinal direction, but there was little to provide transverse stiffness or strength. The upper stories, on the other hand, were stiffened by interior transverse partitions and walls. The soft ground story limited the seismic force transmitted to the upper stories so that they came through relatively undamaged, but the ground story itself was severely distorted and hence was severely damaged in the process. In effect, the soft ground story acted as a structural fuse, albeit a non-expendable one, limiting the force transmitted through it but at the cost of its own destruction.

Lack of redundancy contributed to the collapse of the Bucharest computing center in the Romanian earthquake of 1977 (Ref. 8). The building had been designed to meet the seismic provisions of the Romanian building code, which stipulated a design base shear for this building of 6% of its weight. The ground story columns provided the only resistance to lateral forces. There was no second line of defense. The main building, 30 meters square and three stories high, had service towers at opposite ends, structurally separated from the central building. Figure 24 shows the building, and Figs. 25 and 26 a ground floor layout and a typical column.

The central building was cast-in-place reinforced concrete construction with precast concrete exterior walls above the ground story. Nine columns spaced 12 meters apart supported the entire building. The walls enclosing the ground story detoured around the columns and were capped with a continuous band of windows that separated them structurally from the floor

Figure 24. Computing center, Bucharest, Romania, before the earthquake

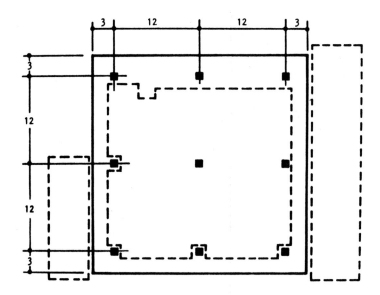

Figure 25. Ground floor layout, Bucharest computing center.

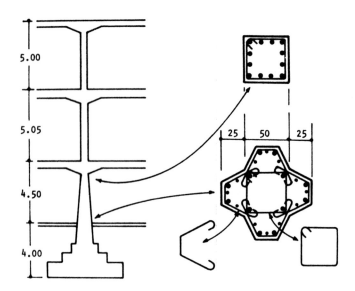

Figure 26. Typical column and slabs, Bucharest computing center.

98

above. There were no shear walls. Thus the only structural elements that could resist lateral forces were the columns, unaided by lateral restraint of any kind. Their failure produced the collapse shown in Fig. 27. The fluted, tapered shape of the ground story columns and the use of hairpin bars instead of closed loops for ties no doubt reduced the shear capacity of the columns. Figure 28 shows the top of one, with its ties burst open and the concrete sheared. When the columns failed, there was nothing else to carry the load and the central building collapsed.

In contrast, consider the 14-story Mt. McKinley Building (Figs. 29 and 30) in Anchorage, damaged in the Alaska earthquake of 1964 (Refs. 19 and 20). Many of the spandrel beams were sheared and some of the main columns were severely damaged, including a pier column in the end wall that was completely severed. The building was of cast-in-place reinforced concrete construction, with plenty of redundancy. When one element failed, other elements were able to pick up the load. Damage was extensive, but repairable.

The lateral forces of the earthquake may exceed the design lateral forces by a substantial margin, as we saw earlier in Fig. 5. For the most part, factors of safety and redundancy provide the needed margins of strength and resilience. But the forces of the earthquake may act in either direction, and therefore their effects may either augment or offset those of gravity or other design loads. Indeed, in some instances they may more than offset them, leading to unanticipated stress reversals.

A case in point is the telecommunications building in Managua, damaged in the Nicaragua earthquake of 1972 (Ref. 21). The structure for the Telex building was a reinforced concrete frame with prestressed girders spanning between exterior columns at second floor level. Figure 31 shows one of the girders. As the building swayed back and forth in the earthquake, the shear force induced in this girder due to frame action oscillated between positive and negative, thereby augmenting gravity load shear at one end of the cycle while offsetting it at the other. This brings to our attention a special problem posed by prestressed girders, in which the strands are usually draped to take part of the shear due to self weight. Here, a part of the gravity load shear in the concrete is already relieved by the draped strands, and the

Figure 27. Collapsed computing center, Bucharest.

Figure 28. Ground story column failure, Bucharest computing center.

Figure 29. Mt. McKinley Building, Anchorage.

Figure 30. End view of Mt. McKinley Building, Anchorage.

Figure 31. Prestressed girder in Telecommunications Building, Managua, Nicaragua.

earthquake response may produce a stress reversal not taken into consideration in the design of the member. Note that the shear crack in the girder shown in Fig. 31 slopes downward toward midspan, whereas a gravity load shear crack would slope downward toward the end.

The corner columns of tall buildings may be particularly vulnerable to seismic overloads or stress reversals, since the overturning moment primarily affects the exterior columns, and the corner columns are exterior to both directions of response. American building codes now in force do not require a consideration of seismic forces acting concurrently in two directions, although the ATC 3-06 provisions do, and many engineers routinely take concurrent action into account.

A corner column in one of the Karpos Tower apartment buildings (Fig. 32) in Skopje, Yugoslavia, failed in compression in the 1963 Skopje earthquake, whereas the other columns in the building showed little distress (Ref. 22). In the 21-story Petunia II apartment building (Fig. 33) in Caracas, Venezuela, damaged in the 1969 earthquake, the axial force in one corner column due to overturning moment more than offset that due to gravity, producing tension cracks in the column (Refs. 17 and 23).

Koyna Dam in India, which suffered earthquake damage in 1967, provides another example of unexpected stress reversal (Fig. 34). It is a straight concrete gravity dam with an unusual break in the slope of the downstream face, the result of a design change which had been made after construction had already begun, in order to permit building the dam to its final height in a single stage instead of stopping the first stage at a lower height as originally planned and then adding to its height later in a second stage. Theoretically, under normal service loading, including code seismic forces as well as hydrostatic forces, the vertical stresses in the concrete would be compressive throughout the entire dam for all possible reservoir levels. With the reservoir full, the compression would be small in the upstream face, especially opposite the break in the downstream face, but it would still be compression. In the actual earthquake, however, tension fracture occurred at the upstream face, requiring extensive repairs (Ref. 17).

Figure 32. Corner column failure in Karpos apartment building, Skopje, Yugoslavia.

Figure 33. Corner column failure, Petunia II apartment building, Caracas, Venezuela.

Figure 34. Koyna Dam, India.

The Koyna earthquake was recorded in the dam. The vertical stress in the upstream face due to gravity plus the computed first mode of response to the recorded earthquake is shown in Fig. 35, along with that due to gravity plus code seismic forces. Although code stresses would be compressive, the computed tensile stresses in the upstream face far exceeded the tensile strength of the unreinforced concrete.

Connections are the most complex parts of a structure and often are points of weakness. A cement bin (Fig. 36) in Anchorage, which collapsed in the Alaska earthquake of 1964, had a substantial supporting structure with diagonal bracing between the columns. The column bases were quite adequate for vertical loads, but the welded connection of the columns and diagonal braces to the bearing plates failed, as shown in Fig. 37. The connection was nominal at best.

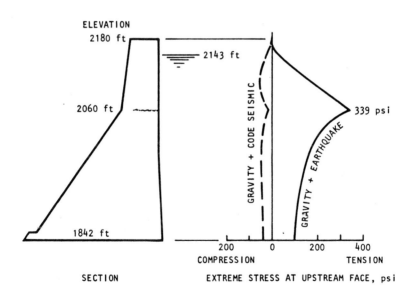

Figure 35. Extreme stresses in upstream face of Koyna Dam.

Partial collapse of the Alaska Sales and Service Building (Fig. 38) in Anchorage in the same earthquake also resulted from inadequate connections, in this case the connections joining

109

Figure 36. Cement bin, Anchorage.

110

Figure 37. Column base plate for cement bin, Anchorage.

precast concrete roof tees to the supporting hammerhead columns, and joining the hammerhead columns end to end. This building was not quite complete when the earthquake struck. It was rebuilt and strengthened, using most of the original precast elements, but this time the connections were modified to provide far greater resistance to lateral forces.

Brittle nonstructural materials should be used warily. Banco Central in Managua was a reinforced concrete frame building with clay tile filler walls (Fig. 39). As the building swayed in the 1972 earthquake, the brittle filler walls were badly cracked. Although the structure itself was not badly distorted, nonstructural breakage was extensive. The clay tile walls that enclosed the stairwells shattered, leaving debris scattered down the stairs as shown in Fig. 40. Fortunately, the building was unoccupied when the earthquake struck. Had it been necessary for workers to try to get out of the building down the debris-littered stairs — in a panic — without lights — the injuries might well have been appalling.

It can never be known how much earthquake damage has been prevented by compliance with seismic building codes. Opportunities for comparative evaluation are few, but the Kern County, California, earthquake of July 21, 1952, provided one. Prior to 1933, school buildings in California were designed without specific seismic regulations. Following the 1933 Long Beach earthquake, the California State Legislature passed the Field Act, which gave the Division of Architecture responsibility for regulating the design and construction of public schools throughout the state. Seismic design and construction provisions were adopted, as well as rather strict inspection requirements. Existing buildings were not modified, but new school buildings were subject to the new requirements.

Within the region of destruction of the 1952 earthquake were school buildings built before the imposition of seismic code requirements and after. Steinbrugge and Moran, after an extensive field study of the effects of the earthquake, reported that school buildings constructed under the controls of the Field Act were practically undamaged, whereas the older school buildings not subject to the Field Act were seriously affected (Ref. 24). The age factor was probably not significant, for all of the buildings

Figure 38. Alaska Sales and Service building, Anchorage.

113

Figure 39. Banco Central, Managua, Nicaragua.

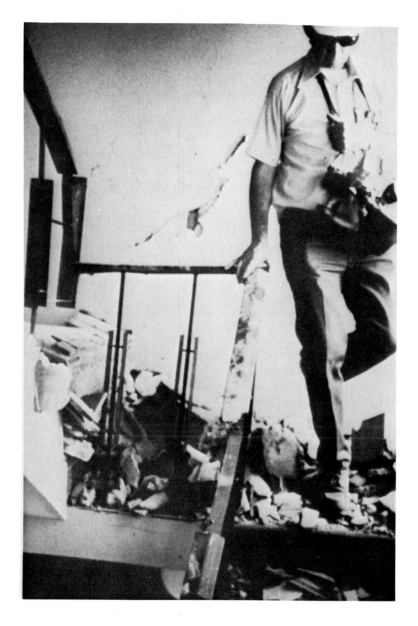

Courtesy of Henry J. Degenkolb

Figure 40. Debris-littered stairway, Banco Central, Managua.

115

were in regular use and had been properly maintained. A tabulation of the effects of the earthquake on 51 masonry buildings at 26 public schools in Kern County describes each building, the degree of damage to it, and whether it was or was not built in compliance with the Field Act. The results show:

Number of School Buildings by Degree of Damage

Damage	Conform to Field Act	
	Yes	No
None	11	1
Slight to minor	6	9
Moderate to considerable	1	9
Severe	0	13
Collapse	0	1

The dramatic difference confirms that the Field Act was effective in reducing the earthquake hazard. That some buildings conforming to the act were damaged suggests that the regulations were not too severe.

Seismic building codes have generally been developed without detailed analyses of the economic benefits and costs, and their cost effectiveness remains vague. The incremental cost of design and construction imposed by seismic regulations would be difficult to determine: A few percent might be a reasonable speculation — a small price to pay for protection if the earthquake occurs. However, even in the most active seismic regions of the United States, a building is unlikely to be subjected to the forces of a strong earthquake during its intended useful life. On the other hand, better earthquake resistance means better resistance to other forces of nature as well. The level at which the incremental cost of augmented earthquake resistance would exceed the incremental benefit is far from clear.

We cannot reasonably expect to be able to forecast earthquakes accurately, much less prevent them, in the foreseeable future. Lacking that ability, we can only try to limit their destruction by designing our buildings to withstand them. Compliance with seismic building codes and heed to the principles of structural dynamics and the lessons of past earthquakes are the best tools we have.

References

1. American Iron and Steel Institute, *The Agadir, Morocco Earthquake*, New York, 1962.

2. Wegener, A., *Die Entstehung der Kontinente und Ozeane*, 4th ed., Braunschweig, Friedr. Vieweg & Sohn, 1929.

3. Chopra, A., *Dynamics of Structures, A Primer*, Earthquake Engineering Research Institute, Berkeley, California, 1981.

4. Tanabashi, R., "Earthquake Resistance of Traditional Japanese Wooden Structures," *Proc. 2nd World Conf. on Earthquake Engineering*, Tokyo, 1960.

5. Freeman, J.R., *Earthquake Damage and Earthquake Insurance*, McGraw-Hill, New York, 1932.

6. Applied Technology Council, *Tentative Provisions for the Development of Seismic Regulations for Buildings*, ATC 3-06, National Bureau of Standards, Washington D.C., 1978.

7. International Conference of Building Officials, *Uniform Building Code, 1982*, Whittier, California, 1982.

8. Berg, G. V.; Bolt, B.A.; Sozen, M.A.; and Rojahn, C., *Earthquake in Romania March 4, 1977: An Engineering Report*, National Academy Press, Washington D.C., 1980.

9. Building Officials and Code Administrators International, *The BOCA Basic Building Code - 1981*, Homewood, Illinois, 1981.

10. American Insurance Association, *The National Building Code 1976*, New York, 1976.

11. Southern Building Code Congress International, *Standard Building Code, 1976 Edition*, Birmingham, Alabama, 1976.

12. American National Standards Institute, *American National Standard Building Code Requirements for Minimum Design Loads in Buildings and Other Structures* (ANSI A58.1-1972), New York, 1972.

13. American Institute of Steel Construction, *Specification for the Design, Fabrication and Erection of Structural Steel for Buildings*, New York, 1978.

14. American Concrete Institute, *Building Code Requirements for Reinforced Concrete* (ACI 318-71), Detroit, 1971.

15. International Conference of Building Officials. *Uniform Building Code Standard 23.1, 1979: Determination of the Characteristic Site Period T_s*, Whittier, California, 1979.

16. Richter, C. F., *Elementary Seismology*, W. H. Freeman and Co., San Francisco, 1958.

17. Berg, G.V. and Hanson, R.D., "Engineering Lessons Taught by Earthquakes," *Proc. 5th World Conf. on Earthquake Engineering*, Rome, 1973.

18. Berg, G.V. and Husid, R., "Structural Effects of the Peru Earthquake," *Bull. Seis. Soc. Amer.*, 61:3, Jun 1971, pp. 613-31.

19. Berg, G.V. and Stratta, J.L., *Anchorage and the Alaska Earthquake of March 27, 1964*, American Iron and Steel Institute, New York, 1964.

20. Berg, G.V., "Response of Buildings in Anchorage," *The Great Alaska Earthquake of 1964* (8 vols.), Engineering vol., National Academy of Sciences, Washington D.C., pp. 247-82.

21. Berg, G.V. and Degenkolb, H.J., "Engineering Lessons from the Managua Earthquake," *Managua, Nicaragua Earthquake of December 23, 1972* (2 vols.), Vol. 2, Earthquake Engineering Research Institute, Oakland, California, 1973, pp. 746-67.

22. Berg, G.V., *The Skopje, Yugoslavia Earthquake, July 26, 1963*, American Iron and Steel Institute, New York, 1964.

23. Hanson, R.D. and Degenkolb, H.J., *The Venezuela Earthquake, July 29, 1967*, American Iron and Steel Institute, New York, 1969.

24. Steinbrugge, K.V. and Moran, D.F., "An Engineering Study of the Southern California Earthquake of July 21, 1952, and Its Aftershocks," *Bull. Seis. Soc. Amer.*, 44:2b, Apr 1954, pp. 199-462.

119